P9-DBK-922

Calculators in Mathematics Education

1992 Yearbook

James T. Fey
1992 Yearbook Editor
University of Maryland

Christian R. Hirsch
General Yearbook Editor
Western Michigan University

National Council of Teachers of Mathematics

Library of Congress Cataloging-in-Publication Data:

Calculators in mathematics education / James Fey, 1992 yearbook
editor, Christian R. Hirsch, general yearbook editor.
p. cm. — (Yearbook : 1992)
Includes bibliographical references.
ISBN 0-87353-342-9 : $18.00
1. Mathematics—Study and teaching. 2. Calculators. I. Fey,
James Taylor. II. Hirsch, Christian R. III. National Council of
Teachers of Mathematics. IV. Series: Yearbook (National Council of
Teachers of Mathematics) : 1992.
QA20.C34C35 1992
510'.71'2—dc20 91-47745
CIP

The publications of the National Council of Teachers of Mathematics present a variety of viewpoints. The views expressed or implied in this publication, unless otherwise noted, should not be interpreted as official positions of the Council.

Printed in the United States of America

Contents

Part 3: Calculators in 9–14 Mathematics

Part 4: Calculators in Mathematics Assessment

Part 5: Strategies for the Implementation of Calculators

Part 6: Calculator Activities for the Classroom

Preface

It is now nearly twenty years since the first hand-held electronic calculators were offered for sale as everyday consumer products. Not surprisingly, these first calculators fascinated mathematics students and teachers. They prompted spirited debate about possible changes in the goals and teaching approaches of school mathematics—especially elementary school arithmetic. However, the pace of calculator-induced change in school curricula has been very slow.

When microcomputers burst on the mathematics education scene in the early 1980s, they quickly overshadowed interest in the potential uses of calculators. But recent developments in calculator technology—including capabilities for the manipulation of fractions, for graphing algebraic expressions, for statistical data analysis and display, and for formal symbolic manipulation—have brought low-cost, hand-held, battery-powered devices back to center stage in the debate over new directions for school mathematics. The new vision of school mathematics expressed in the NCTM *Curriculum and Evaluation Standards* and *The Professional Teaching Standards* is predicated in part on advances in calculator technology and the assumption that appropriate calculators will be available to all students at all times.

This NCTM 1992 Yearbook has been designed to provide a stimulating collection of articles on the impact and potential of calculators. We sought contributions that examined traditional assumptions about the content and organization of mathematics courses and approaches to teaching and the assessment of student learning. We invited articles that described successful classroom practices that used calculators for teaching mathematical ideas and procedures and problem solving. We also looked for articles that described successful strategies for implementing calculators in classes, schools, and systems. More than 85 proposals were submitted to the initial panel review for the yearbook, and the final publication has 36 contributions from 47 authors.

The yearbook articles are organized into six major sections. The first section includes four articles that identify the most promising and important ways that calculators can influence both the content and the process of mathematics instruction in grades K–14 and a summary of research evidence that can be used in judging the available options. The second section includes five articles describing ways that calculators have been used successfully for mathematics instruction in the primary, elementary, and middle grades. The third section includes seven articles that address calculator use in high school and college mathematics teaching—from algebra and statistics to calculus and discrete mathematics. The fourth section contains four reports of ways that calculators have influenced the testing in state assessment programs and college admissions. The fifth section includes five reports of successful strategies for implementing calculator use in classrooms, schools, school districts, and entire states, and the sixth section contains descriptions of classroom calculator activities that illustrate some of the new instructional possibilities.

The production of this yearbook has been made possible only by the thoughtful contributions of many people. The editorial panel played a critical role in shaping the call for papers and in reviewing the proposals and first drafts of all manuscripts. This demanding work and careful decision making was shared by the following people:

Nancy Cetorelli	Stratford, Connecticut, Public Schools
Christian Hirsch	Western Michigan University
John Kenelly	Clemson University
Thomas Rowan	Montgomery County, Maryland, Public Schools
Sheila Sconiers	University of Chicago School Mathematics Project

Throughout the yearbook production process, from planning meetings to the final editing of manuscripts and preparation for printing, the editors and the panel have been supported well by the outstanding NCTM headquarters staff.

In all this work, special credit must go to Chris Hirsch, the general editor of the past several NCTM yearbooks. Chris conceived the idea of a yearbook on calculators at a time when few in mathematics education could anticipate the dramatic potential of graphing calculators, and he provided wise and provocative counsel at every stage of development and editing.

We hope that you will find a combination of stimulating and practical ideas in the yearbook and that the years ahead will see further creative and practical developments applying the power of calculators to the improvement of school mathematics.

JAMES T. FEY
1992 Yearbook Editor

1

Part 1

The Potential for Calculators to Transform Elementary School Mathematics

Grayson H. Wheatley
Richard Shumway

CALCULATORS have the potential to transform school mathematics from a procedurally dominated subject to the exciting study of patterns and relationships. This can happen when we understand what it is like to study mathematics with the calculator as a tool and when we can use instructional materials that assume calculators are widely available. Although inexpensive calculators have been available for more than a dozen years, few elementary school teachers use them (Reys et al. 1990).

The use of calculators in elementary school mathematics cannot be considered without addressing issues related to teaching paper-and-pencil computations. What level of paper-and-pencil computational proficiency is appropriate in today's society? How important is it that students be able to divide 3495 by 531, when this task can be done using a calculator in a few seconds? By what process do students decide whether to use a calculator, to estimate, to compute mentally, or to use paper and pencil?

ABANDON THE TEACHING OF COMPUTATIONAL PROCEDURES

In a 1990 report entitled *Reshaping School Mathematics,* a National Research Council committee recommended a "zero based" approach to school mathematics curriculum development. Although we must be careful using corporate metaphors, the "zero based" metaphor may have merit in helping us break out of our view of school mathematics determined by "the way things were" to think about "what could be." In this article we recommend the construction of school mathematics without taking computational procedures as an assumed component. We believe that a curriculum could be devised in which calculators would be used for all but 15 percent of school mathematics. We would not "teach" paper-and-pencil procedures for adding and subtracting three-digit numbers in grades 2 and 3, the long-division algorithm in grades 4, 5, and 6, nor the addition and subtraction of fractions in grades 5–8.

Arithmetic would have a place in our "zero based" curriculum, but it would have new goals and emphases. It would be much more important that students *know when to subtract* than that they be able to use a prescribed and complex subtraction algorithm efficiently. Mathematics would be characterized by the search for patterns and relationships rather than fixed procedures to be mastered (Steen 1990). With the use of calculators, attention would focus on meaning, and mathematics would become a much more exciting activity for students. Mental arithmetic and estimation would become major components of the school mathematics curriculum. Students would be encouraged to create their own algorithms for simple computations, but they would be encouraged to use calculators whenever it made sense to do so.

Viewed in the long tradition of tool making, calculators represent a powerful alternative to the drudgery and inefficiency of paper-and-pencil arithmetical computation. Arguments that learning the school arithmetic algorithms contributes to mathematical knowledge ring hollow. On the contrary, the effects of practicing poorly understood procedures may actually hinder learning to act mathematically. It is not at all clear that becoming proficient with taught computational procedures is central to acquiring mathematical thinking and problem-solving power, even though most of school mathematics has that goal. In fact, learning and practicing the long-division algorithm may foster the development of beliefs that "understanding" a procedure is understanding mathematics (Wheatley 1980). Just remembering a set of steps in computing is counter to constructing number relationships. It is indeed more likely that calculator-enhanced activities designed to facilitate the construction of mathematical relationships (illustrated below) would be more appropriate and effective.

CALCULATORS AND NUMBER DEVELOPMENT

The calculator can play an important role in young children's number development. In fact, using a calculator to perform multidigit subtraction has many advantages over trying to use any of the complex paper-and-pencil algorithms. Grade 3 children have great difficulty subtracting with three-digit numerals. The difficulty lies in constructing 10 and 100 as abstract units (Cobb and Wheatley 1988) and in coordinating units of three different ranks (1, 10, 100). Unless children have made these constructions, it is unlikely they can meaningfully perform three-digit subtraction using the standard algorithm. Recognizing the cognitive difficulty of this task, textbook authors have provided a method that allows answers to be obtained without relational understanding (Skemp 1978). Although an attempt has been made to help students understand the procedure, this is not the same as carrying out the cognitive operations necessary to give meaning to the mathematics. However, the focus on "understanding" this algorithm may foster the belief that mathematics *is* procedures—someone else's procedures.

Students can give meaning to three-digit numbers long before they can meaningfully use the three-digit algorithm of school mathematics. For evidence, see the article by Shuard in this yearbook. If a calculator is used to subtract 386 from 517, students can solve problems using large numbers and in the process make constructions that will lead to mathematical power. They can develop and use number sense without being frustrated by an unreasonable demand, that is, using a complex procedure developed for a different era. Thus, calculator computing can facilitate the construction of mathematical relationships and result in greater mathematics learning than training students to use a procedure at an age when many cannot possibly make sense of what they are doing.

When mathematics learning is characterized by meaningfulness, mental arithmetic becomes an inherent part of doing mathematics. Informal and practical computational procedures can be constructed by the learners, and the distinction between conceptual and procedural knowledge (Hiebert 1986) may not be useful when mathematics is viewed as constructing meaning.

Number sense is a major theme of the NCTM *Curriculum and Evaluation Standards for School Mathematics* (1989). Since number sense requires a network of interrelated conceptual schemes and operations, it is important that students give meaning to mathematical ideas by constructing the objects of their personal mathematical worlds (Wheatley 1991). For example, a child who can confidently determine 10 more than 73 has constructed 10 as an iterable unit, a major construction that is essential in building number sense (Cobb and Wheatley 1988). At this point calculators can contribute to building number sense, as illustrated by the following activity.

A CALCULATOR-ENHANCED MATHEMATICAL ACTIVITY

In the Range Game (Wheatley 1990) the class is asked to enter 75 in a calculator and choose a number that, when added to 75, gives a number in the range from 90 to 110. As the class suggests possible numbers, the teacher writes them on the board. Questions about the smallest or largest number naturally arise in this discussion, and soon the list of whole numbers from 15 to 35 is obtained. Following is the dialogue that resulted when a Multiplication Range Game was played with grade 6 students.

Teacher: Find a number that when multiplied by 37 gives a number in the range from 500 to 600.

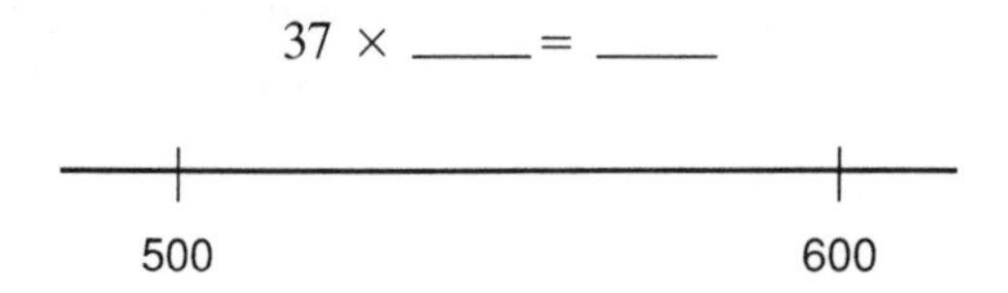

The teacher writes the numbers suggested by students on the board.

Sandy: 15.

Pete: 18.

Faith: 14.

Wendy: 18 doesn't work [teacher crosses off 18 without comment].

Mike: 16 is the largest number.

Teacher: Why do you say 16 is the largest?

Mike: Because I tried 17 in my calculator, and it was too big.

Eddie: 14 is the smallest.

Teacher: Do we have them all now?

Class: Yes.

Ann: Wait a minute. 14 is *not* the smallest. I tried 13.9 in my calculator, and it works (75 × 13.9 = 514.3).

This unexpected response launched the class into a fervent search for more "numbers like that."

The Range Game embodies computational estimation, decimals, and limits and in general promotes number sense. With computations performed on a calculator, the student can focus on number relationships. There are many other activities where the calculator plays an instrumental role in developing mathematical reasoning and concepts (see Reys et al. 1979).

CALCULATORS AND PROBLEM SOLVING

Calculators have a particularly beneficial effect when used in mathematical problem solving. Hembree and Dessart's research meta-analysis (see their article in this yearbook) showed that problem-solving scores are higher when calculators are used. Students are more effective in selecting and using appropriate problem-solving heuristics and are more confident and enthusiastic when allowed to use calculators in solving mathematics problems (Hembree and Dessart 1986; Reys and Reys 1987; Szetela and Super 1987; Wheatley and Wheatley 1982). Thus, calculators seem to be particularly useful when students are "doing mathematics."

Understanding the Problem

Calculators can be quite useful in deciding *how to solve* a problem. In the act of devising a problem-solving plan, good problem solvers will often perform a variety of exploratory moves—they will try certain computations in the process of *getting to know* the problem.

Consider this problem: "Notebooks cost $1.57 each. How many notebooks can Pat buy for $40.00, and how much change will she receive?" As an exploratory move, grade 6 pupils equipped with calculators have been observed trying 10 × $1.57 and then 15 × $1.57 and then 22 × $1.57 in the process of deciding how to solve the problem. These exploratory moves are greatly facilitated by calculators. Such exploratory methods were not observed as frequently in students who had not had classroom calculator experience, even though a calculator was available at the time. Thinking of doing the computations "by hand," many of these students would do nothing rather than carry out the laborious computations; the calculator-experienced students, however, readily tried a variety of computations in their attempt to understand the problem and devise a plan (Wheatley and Wheatley 1982). Computational tools can free students to reason mathematically without being sidetracked by carrying out time-consuming paper-and-pencil algorithms.

Deciding on a Method

When calculators are introduced in the classroom, students do not spontaneously use them, even for complex computations. It is necessary for individuals to construct action schemes that include calculator use as a viable option. Sixth-grade students, without extensive calculator experience, have been observed performing long division by paper and pencil when a calculator was lying on their desk (Wheatley and Wheatley 1982). In some cases attempts to use a calculator were rejected because the students were unable to give meaning to the resulting decimal display. It was clear they had been expecting a whole number and possibly a remainder as with paper-and-pencil computations and were unprepared for the decimal that appeared.

When an individual makes the decision to use a calculator in some way, she or he often performs a *thought experiment* (Lakatos 1976). In deciding how to carry out certain arithmetical operations, an individual will often construct an anticipated sequence of moves and "run through" the activity mentally before actually entering numbers. Such thought experiments allow an individual to explore and evaluate the efficacy of using a calculator in comparison to other possible methods. The question becomes, Should I perform this calculation mentally? Estimate? Use paper and pencil? Use a calculator? Discuss it with someone? Use a computer? In order to become powerful users of calculators, students must form how-to-proceed schemes through wide experience with such thought experiments.

Creating Problematic Situations

Calculators can be used to create problematic situations. For example, first ask students to enter 53 [+] 20 [=] 34 [=] 50 [=] . . . and then ask, "What is happening? What is the calculator doing?" This task encourages students to look for patterns and put forth conjectures that they can then test. Many four-function calculators will display 73, 54, and 70; others will display 73, 87, and 103. This activity provides opportunities for students to give meaning to the significance of the order of addends and the order of operations in computations. This activity could be followed with the question, "Does your calculator give the same result for $7 \times 8 + 10$ as it does for $10 + 7 \times 8$? How can you explain this?" Three calculators we tested gave 66, and three others gave 136.

Calculators provide excellent opportunities for students to construct meaning for integers. Having students count back from 10 using a built-in constant leads eventually to 0, -1, -2, -3, -4, -5, -6, ... Understanding rational numbers and operations on them can also be enhanced by the use of calculators. Interesting number patterns are generated in division. For example, $4/11 = 0.3636363$ and $7/11 = 0.6363636$. Are these the same? What is going on here? Many other intriguing results can be obtained that can pique the interest of students and enhance understanding. Meaning for decimals can be developed by such activities as this: Ask students to divide 3 by 8, giving 0.375. Challenge them to find a number that, when added to 0.375, yields 0.385. The meaning of hundredths can grow from such experiences.

Persistence

One factor that contributes to success in problem solving is persistence. Do students work at mathematics tasks for extended periods of time until a solution is reached, or do they quit when the first obstacle is encountered? Students in conventional instruction often develop the belief that any problem should be solved in less than a minute. Of course, most significant mathematics questions require considerable thought and reflection. Belief that problems should be quickly solvable can be a debilitating factor in students' problem-solving performance, but there is evidence that when students are engaged in problem solving with calculators, they become more persistent—they will work at tasks for long periods of time. The calculator appears to play a significant role in this change (Hembree and Dessart 1986; Wheatley and Wheatley 1982).

BEYOND THE FOUR-FUNCTION CALCULATOR

Fractions are not well understood by many students. This problem may result from emphasis on procedures for computing with numbers in fraction form, which may tend to obscure conceptual understanding. Being able to use fractional and decimal forms is a universally accepted goal that is not being realized. Several low-cost calculators have the capability of performing computations with numbers in fraction form, displaying the answer as a fraction, simplifying fractions, and converting fractions to decimals. By considering tasks for which fractional notation is useful, students can build meaning for fractions and mathematical relationships rather than practicing poorly understood computational rules. Using a fraction-friendly calculator, rather than teaching the least common denominator method for adding fractions with unlike denominators, can focus attention on constructing fractions like 2/3 as mathematical objects that are part of a larger schema including other types of numbers. As with all tools, using fraction calculators extends the conceptual reach of our students.

Although graphing calculators are used primarily in high school mathematics, their use at lower grades also has great potential (see the article by Vonder Embse in this yearbook). Elementary and middle school students can give meaning to variables, functional relationships, and equations. By using a graphing calculator, students can come to associate a visual image with symbolic expressions. Both linear and nonlinear relationships can be conceptualized when the capability exists to generate graphs of these functions with a few keystrokes (Shumway 1992). Students can come to think dynamically about mathematical equations when they can watch curves being drawn right in front of their eyes and when they can easily repeat the graphing as many times as necessary.

CONCLUSION

The fear that calculators will undermine mathematics learning is widespread, in spite of the fact that no evidence to support such a position exists. On the contrary, considerable research evidence suggests that calculators can play an important role in children's mathematics learning (Hembree and Dessart 1986; Wheatley et al. 1979). In an environment where calculators are available, the emphasis of mathematics instruction can be on meaningful concept development and problem solving and not on computational procedures. We believe that such a curriculum is both appropriate and long overdue, and we urge all mathematics educators concerned with elementary school curricula and teaching to work on the development and implementation of programs that realize the great promise.

REFERENCES

Cobb, Paul, and Grayson Wheatley, "Children's Initial Understanding of Ten." *Focus on Learning Problems in Mathematics* 10 (1988): 1–28.

Hiebert, James. *Conceptual and Procedural Knowledge: The Case of Mathematics.* Hillsdale, N.J.: Lawrence Erlbaum Associates, 1986.

Hembree, Ray, and Donald Dessart. "Effects of Hand-held Calculators in Precollege Mathematics Education: A Meta-Analysis." *Journal for Research in Mathematics Education* 17 (March 1986): 83–99.

Lakatos, Imre. *Proofs and Refutations: The Logic of Mathematical Discovery.* Cambridge: Cambridge University Press, 1976.

National Council of Teachers of Mathematics. *Curriculum and Evaluation Standards for School Mathematics.* Reston, Va.: The Council, 1989.

National Research Council. *Reshaping School Mathematics: A Philosophy and Framework for Curriculum.* Washington, D.C.: National Academy Press, 1990.

Reys, Barbara, and Robert Reys. "Calculators in the Classroom: How Can We Make It Happen?" *Arithmetic Teacher* 34 (February 1987): 12–14.

Reys, Robert, Barbara Bestgen, Terrence Coburn, Harold Schoen, Richard Shumway, Charlotte Wheatley, Grayson Wheatley, and Arthur White. *Keystrokes: Calculator Activities for Young Students* (four booklets). Palo Alto, Calif.: Creative Publications, 1979.

Reys, Robert, Barbara Reys, Rita Barger, J. Hauck, L. Morton, S. Reehm, R. Sturdevant, and J. Wyatt. *Calculator Use in Mathematics Teaching in Missouri Schools: A 1990 Status Report.* Columbia, Mo.: University of Missouri, 1990.

Skemp, Richard J. "Relational Understanding and Instrumental Understanding." *Arithmetic Teacher* 26 (November 1978): 9–15.

Shumway, Richard J. "Calculators and Computers." In *Teaching Mathematics in Grades K–8: Research-based Methods,* 2d ed., edited by Thomas Post. Boston: Allyn & Bacon, 1992.

Steen, Lynn Arthur. *On the Shoulders of Giants: New Approaches to Numeracy.* Washington, D.C.: National Academy Press, 1990.

Szetela, Walter, and Doug Super. "Calculators and Instruction in Problem Solving in Grade 7." *Journal for Research in Mathematics Education* 18 (March 1987): 215–29.

Wheatley, Grayson. "Calculators and Constructivism." *Arithmetic Teacher* 38 (October 1990): 22–23.

———. "Calculators in the Classroom: A Proposal for Curriculum Change." *Arithmetic Teacher* 28 (December 1980): 37–39.

———. "Constructivist Perspectives on Mathematics and Science Learning." *Science Education* 75 (January 1991): 9–21.

Wheatley, Grayson, Richard Shumway, Terrence Coburn, Robert Reys, Harold Schoen, and Arthur White. "Calculators in Elementary Schools." *Arithmetic Teacher* 27 (September 1979): 18–21.

Wheatley, Grayson, and Charlotte Wheatley. *Calculator Use and Problem Solving Strategies of Grade Six Pupils: Final Report.* (ERIC Documents Number ED 175720.) Washington, D.C.: National Science Foundation, 1982.

2

Calculators in the Middle Grades: Access to Rich Mathematics

Nicholas A. Branca
Becky Ann Breedlove
Byron W. King

CALCULATORS play a prominent role in the model classroom envisioned by recent reform proposals in mathematics education. Recommendations like the NCTM *Curriculum and Evaluation Standards for School Mathematics* (1989), *Everybody Counts* (1990), and *Reshaping School Mathematics* (1990) all advocate rich mathematical experiences for *all students*. This notion is especially crucial at the middle and secondary school levels where students are now commonly sorted, according to judgments of their ability and prior achievement, into multiple course sequences that furnish quite different levels of mathematical challenge and growth.

The reform proposals suggest that instead of sorting students into ability and achievement levels, middle school mathematics should engage all students in a varied curriculum that prepares them well for what will be offered in high school. Central to this perspective is the notion that equity and access in mathematics need to be placed prominently at the forefront of educational reform. Calculators are an important tool to overcome obstacles that have prevented many students from having full access to courses rich in mathematical content. The NCTM *Standards* (1989) suggests that for middle grades calculators, along with other technology, should be available to "free students from tedious computations and allow them to concentrate on problem solving and other important content." The *Standards* assumes that all students will have calculators to do their work.

At O'Farrell Community School, where two of the authors teach and the third author works as a consultant, we have tested the hypothesis that calculators can help students surmount traditional barriers and experience success in a strong middle grades mathematics program. O'Farrell is an ethnically diverse urban middle school that enrolls all students in a single course of study. All sixth and seventh graders are in a prealgebra course, and all eighth graders enroll in algebra. Some of the students have poor computational skills, some have special educational needs, some speak English as a second language, and some have been identified as gifted. Our experience with these students has greatly expanded our perspective on the use of calculators. The potential value of the calculators comes through in the voices of our students.

> "Calculators help me learn the mechanics of problems."—Gene Harrison, Grade 8

QUESTIONS FOR TEACHERS

Every day teachers decide whether or not to allow students to use calculators in their classrooms. The decision ought to be a conscious action, based on teacher reflection about the goals of instruction. We believe that answers to five key questions should guide calculator use:

1. *Does the calculator allow the students to get closer to mathematical concepts being presented?*

> "Calculators could help you in your math class if you knew the process but just wanted to get your work done faster."—Melissa Camino, Grade 8

Calculators can often be used to help students understand mathematical concepts better. For example, we have found that calculators allow students to understand the nature of exponents without engaging in endless, repetitive multiplications. Students are much more likely to explore the effects of raising several different numbers to the tenth power and find interesting patterns in those results if they have calculators.

Many teachers accept the futility of continuing to teach heretofore unsuccessful middle level students the algorithm for long divison using pencil-and-paper methods. They see little purpose in lowering their students' self-esteem by reminding them of previous failure. But using calculators to perform required division operations can free students to focus on the concepts of division required in problem solving. Calculators can also help students explore the connection between division with whole number remainders and divison in which remainders are reported as common fractions or decimals.

Of course, there are some mathematical topics for which calculators do not seem to offer significant aid in concept development. For instance, we have found that developing an understanding of fraction concepts and operations is most effective through the use of concrete materials. Calculators that are capable of representing fractional expressions are useful but not the most effective focal point of rational number instruction.

Part of the appeal of concrete materials is the fact that many middle school students learn best through tactile activity. We've found that for some of those students the act of touching calculator keys seems to cause a more intense focus on the problems at hand. In fact, manipulating calculator keys becomes the gateway for engaging some students in thinking about mathematical problems and the concepts underlying those problems.

2. *Will the use of the calculator in a mathematics activity increase student confidence and persistence?*

> "Calculators help me to not get so confused with math when I'm working with very high numbers. It helps me relax my mind a little bit."—Jit Panhwanh, Grade 8

Students will not learn mathematics if they refuse to try mathematics. Using calculators lets some students function confidently in difficult classes, despite previous negative experiences. For instance, students in middle grades are often taught that an average is always determined by adding and dividing. Little or no mention is made of median and mode, and typical problems involve only a few data points from situations that are related to real life. With a calculator in hand, children in the middle grades can explore meaningful concepts and problems in statistics with confidence and persistence. Our students amaze us with the amount of data they are now willing to consider and manipulate. During one unit project that incorporates linear measurement and measures of central tendency, students are easily motivated to find the mean, median, and mode of the heights of their classmates, teachers in their school, or a favorite basketball team. When allowed access to a calculator, they enjoy speculating about how answers change when outliers are introduced or removed from the data.

3. *Could the concept be taught with an inductive approach?*

> "Calculators can help us by figuring out how to do difficult problems or confusing numbers."—Henry Lopez, Grade 8

One of the topics of middle-grade mathematics is properties of operations in the number systems. Using calculators, students are able to discover those properties inductively, gaining personal ownership of the mathematics in the process. For example, one of the crucial steps in operations with decimals (even in estimation) is placing the decimal point correctly in the sum, difference, product, or quotient. Though many children in the middle grades can accurately memorize the rules for lining up decimal points, appending zeroes, or counting decimal places in the factors of a product, many others confuse the various rules, and few can tell why the procedural steps are required.

With calculators available, students can approach the learning of rules for decimal operations like real mathematicians. They can start with a blank piece of paper, generate a wide variety of decimal computation problems, use the calculator to get the answers, and compare notes with other students to find and justify patterns observed in the results. Similar investigations can lead students to discover the algebraic order of operation rules. Approaches to this topic are described in some detail in the essay by Vonder Embse in this volume.

Another opportunity for inductive learning is given by the variety of keys on most standard calculators. Students can test the effects of unfamiliar keys in an effort to discover their functions. Again, students can be asked to make up and record the results of problems they have constructed and then generalize the "thinking" of the calculator after comparing results with other students. We have found this type of activity particularly interesting with the keys on fraction calculators.

4. *Would the use of the calculator facilitate the study of real-life applications?*

> "Calculators help a lot in our math classes because they prepare us for the future. We'll be using calculators in high school, college, and even our jobs. Calculators will prepare us for our jobs in the future."—Joshua Jones, Grade 8

Numbers in real life are often messy. They can be very large or very small. Calculators allow teachers to pull timely and relevant problems about the economy and populations from the front pages of the newspaper. (Students must, of course, understand the meaning of large numbers like millions, billions, and trillions.)

In schools without calculators, mathematics and related disciplines generally avoid having students work with realistic numbers encountered in the occupations to which they aspire. We expect students to learn adult life skills without the one tool—the calculator—that most adults now rely on for quantitative problem solving. Population reports, government budgets, and lawsuit judgments have little in common with most schoolbook numbers. The numbers connected to scientific research, baseball statistics (including players' salaries!), rates of inflation, cost of living, and odds of winning the lottery look more like what we sometimes find on the enrichment pages of our textbooks, to be tackled by the few students who are very proficient with paper-and-pencil computational algorithms.

5. *Will using the calculator allow assessment to be focused on relevant educational objectives?*

> "Calculators help me get the answer to a problem when I don't know it and I still have to explain how I get it."—Randal Bailey, Grade 8

When calculators are used in assessment activities, teachers can focus more easily on measuring their students' growth in conceptual understanding and problem-solving ability. For example, if the goal of a unit is for students to develop skill in solving story problems involving percents, the use of calculators in assessment can reduce computational errors with decimal operations that are peripheral to the unit objectives.

Assessment can be made more open-ended and nonroutine when calculators are available to students. For instance, a test on decimal arithmetic could include tasks like preparing a week's food budget for a family of four, including an actual trip to the market.

CONCLUSION

The greatest challenge facing mathematics educators today is to organize instruction so that it attracts, and develops the abilities of, the greatest number of students possible. This challenge is especially great at the middle school level where mathematics courses so often serve a sorting function that tracks students throughout their secondary schooling. All students deserve the opportunity to learn important ideas that are traditionally encountered only in higher-level classes in tracked schools. The use of calculators in the classroom must be connected to that broader objective—providing all students access to a broad range of mathematical ideas.

Teachers in the middle school mathematics classroom described in the *Curriculum and Evaluation Standards* are sophisticated decision makers. They are clear about their educational objective and are constantly revising their educational activities to maximize the number of students who understand selected concepts. In this context, calculators are not a panacea, but merely one more very important tool in the repertoire of teachers seeking to bring out the best from the most students.

REFERENCES

National Council of Teachers of Mathematics. *Curriculum and Evaluation Standards for School Mathematics.* Reston, Va.: The Council, 1989.

National Research Council. *Everybody Counts: A Report on the Future of Mathematics Education.* Washington, D.C.: National Academy Press, 1989.

———. *Reshaping School Mathematics: A Framework for Curriculum.* Washington, D.C.: National Academy Press, 1990.

3

The Graphing Calculator: A Tool for Change

Gail Burrill

As the ninth-grade algebra teacher was walking around the room observing and helping students, Katie, trying to find the equation of a line given two points, looked up and asked, "Is this really the right equation?" Several weeks later in the same class, Paul and Cheryl struggled to find an equation to describe the number of oranges produced in the following situation: When 20 trees per acre were planted, each tree produced 600 oranges, but as the number of trees increased beyond 20, the yield per tree decreased by 15. Their result was complicated, and they were concerned about its validity. In second-year algebra, Adam and Kim, trying to decide what effect the translation $T_{3,-4}$ has on the parabola $y = x^2$, were arguing whether the signs should be plus or minus in the equation of the new parabola. In calculus the teacher waited, chalk in hand, for students to find displacement, velocity, and acceleration for a projectile motion function as time varied over an interval. In each of these situations, as well as in many others over the past two years, the teacher left the room to find a new tool—a tool that would help Katie, Paul, Cheryl, Adam, Kim, and the calculus students learn mathematics—a graphing calculator.

Katie, trying to find the equation of a line, used the calculator's plotting capability to display the given points and then graphed her equation to check that it did indeed represent the line determined by the two points. Paul and Cheryl, in their problem about orange production, already had a table of values as they calculated what would happen if there were 21 trees per acre, or 22 trees per acre, or 25 trees per acre. They had even constructed a reasonable graph from their table. With the calculator they graphed the equation they believed described the situation and used the trace function to see if the ordered-pair coordinates for points on the graph were the same as those in their table (and a loud and proud "Yes!" resounded from their desk.) Kim and Adam used the calculator to explore their conjectures about the effects of possible translations on their graph. The calculus students used the "y=" menu to enter rules for the displacement, velocity, and acceleration functions and quickly generated a table of values that they used to analyze the motion of the projectile. As a next step, they graphed the motion and were able to observe the behavior they had charted in their tables.

Calculators, particularly easy-to-use graphing calculators, present a dramatic new challenge in teaching mathematics. As suggested in the NCTM *Curriculum and Evaluation Standards,* these tools have changed the very nature of the problems important to mathematics and the methods used to investigate those problems. Calculators change activities in the mathematics classroom, raise questions about the mathematics that should be taught, and suggest issues that must be considered in designing curricula and assessment strategies.

Just as the four-function calculator challenged the role of pencil-and-paper skills in arithmetic and the goals of elementary school mathematics, graphing and programmable calculators are forcing a serious examination of the secondary school curriculum. What are the implications for an everyday mathematics classroom? What should and will change in the mathematics that is being taught and in the teaching and learning activities of mathematics classrooms as calculators become part of the daily routine in algebra or trigonometry or calculus?

SKILLS AND UNDERSTANDINGS FOR CALCULATOR USERS

Calculators today can do much more than the simple arithmetic of whole number, decimal, and common fraction operations. Formulas for mean, standard deviation, correlation coefficient, permutations, combinations, and standard functions such as sine, cosine, logarithm, absolute value, and the greatest integer are built into scientific and graphing calculators and accessed by simple keystrokes. A table of values for a function can be generated easily, and one value for a variable x can be used as input for several different functions at the same time. Programs can be written to perform calculations required by formulas like the quadratic formula and functions like the normal distribution. What does this ability to calculate mean to the classroom?

Scientific and graphing calculators, whether applied in algebra or trigonometry or calculus, require student understanding and skill in the use of algebraic logic for the entry of expressions, function rules, equations, and inequalities. Since calculators generally produce results in decimal form, users must be proficient in writing and interpreting numbers in decimal and scientific notation. Although it has always been important for students to have a well-developed number sense and ability to estimate results, calculator use requires a new aspect of number sense related to decimals and error and accuracy and rounding. Problems in trigonometry involving

$\sqrt{3}/2$ or $\pi/2$ will most commonly appear on the calculator screen as decimal approximations, and it seems reasonable to assume that 0.707 should become as familiar to students as $\sqrt{2}/2$. Students should recognize that 6.499999999 is quite possibly representing the exact result 6.5.

When calculators are used in problem solving and applications of mathematical concepts, they often produce results of bewildering apparent accuracy. What does a slope of -1.85594624886 indicate about the change in speed of a swimmer over the course of a race? Averaging the height of 26 students in a class yields a result of 62.31452483 inches—what do the decimal places mean? What are the ground rules? There is nothing more devastating to students than to use their calculator to solve a quadratic with irrational roots, check the answer in the back of the book, and find the "correct" answer to be $x = (4 + \sqrt{10})/2$, not $x = 3.58113883$ or even $x = (8 + \sqrt{40})/4$. Where and when should students be taught about interpretation and accuracy? These are minor points in the sequence of mathematical ideas but major points in developing student understanding.

Since calculators, like all mechanical devices, must be given very precise instructions, students must understand and become careful in the use of algebraic symbols. They must know the difference between a negative sign and the operation of subtraction and learn to use the appropriate keys. They must understand that squaring a negative number yields a positive result and use parentheses accordingly. Similar understanding and care is required for the use of the $\boxed{1/x}$ and $\boxed{y^x}$ keys, if the power of those calculator options is to be exploited.

Calculators can now draw histograms, graph functions, plot points corresponding to individual ordered pairs, and connect those points in sequence. Functions and individual data points can be plotted on the same screen. Standard trace features then help the calculator user to read the coordinates of the points on a graph in sequence. The zoom feature allows graphs and sets of points to be investigated from different perspectives—close up, far away, or at one end of a domain. Regions of a graph can be shaded, and the shading can be done in different densities to highlight different conditions. A wide variety of functions and relations can also be graphed using parametric equations and polar coordinates.

To use the power of graphing calculators to produce informative graphs of functions and relations, students need skills in algebraic estimation: What domain and range are reasonable for the given function? What are appropriate scales for the axes? Teaching students how to make those judgments is not prominent in the traditional course sequence, and perhaps because of this, many "broken" calculators have been fixed quickly by suggesting an alteration of the viewing window. Again, precision in the use of mathematical notation and vocabulary is required by calculators. As clever as the new tools are, they do not respond well to a maximum of -10 and a minimum of 10 or to a scale of 1 for an interval from -100 to 100. Would a domain of 6.28 to -6.28 and a scale of 3.14 be sensible for a trigonometric graph? For both axes? In all modes? Perhaps these are technical details of little mathematical significance, but they are essential for effective calculator use.

THE IMPACT OF CALCULATORS ON CURRICULUM SCOPE AND SEQUENCE

Discussing the effects of technology on the curriculum, the Mathematical Sciences Education Board (MSEB) of the National Research Council (1990, p. 25) suggests that computers virtually compel a reordering of traditional topics. The MSEB poses two key questions: "What orders yield optimal learning?" and "What is the relation between the stage of introduction and ultimate understanding?" The use of graphing calculators in the classroom raises the same questions. What should students learn in algebra or trigonometry and in what sequence?

With most current graphing calculators, to graph functions students must enter a rule after the "$y =$" prompt. This means that conditions must be written in terms of one variable. In current first-year algebra courses, it is not until a late chapter on graphing, if at all, that students are asked to rewrite given equations in that form.

To understand the graph of a polynomial function, students must have developed some intuition about the degree of a polynomial, its factors, and their relation to the graph. How should the curriculum be adjusted to help students see the relation between the zeros of a polynomial function, the roots of an equation, and the intercepts of the graph? Because calculators can graph one polynomial as easily as another, should students in algebra progress from linear to quadratic to cubic functions? Should students graph exponential functions when they learn rules for manipulating expressions involving exponents?

Another concern is the present focus on the development of algebraic skills such as polynomial arithmetic. In *Reshaping School Mathematics* (National Research Council 1990), the comment is made that "facility in these skills is not an absolute prerequisite either to the use of mathematics or to further study in mathematical based fields" (p. 19). Consider these problems similar to those in popular first-year algebra texts:

Simplify:

- $(3x^3 - 4x^2y^4 + 2x) + (-4x^4 + 2x^3 + 2x^2y^4)$
- $(p^2q^3r^2s)(pqr^3s^2)$
- $4a\sqrt{a^2b} + a\sqrt{a^2b^3} - 5\sqrt{b^3}$

- $\sqrt{\frac{5}{6}} - \sqrt{\frac{6}{5}}$

The calculator does not give much help in working these problems, but is skill in such algebraic manipulations an important goal for school algebra? When in the mathematics most students will encounter do they need to handle nonlinear expressions with four variables? Expressions such as these are not even part of calculus, although simplifying the derivative of $y = ((x^2 - 3)/(x^2 + 3))^{1/2}$ does not seem much more relevant. Is it essential to have students struggle with such complex problems as preparation for the relatively simple exercises they will legitimately encounter? How hard do problems have to get before students understand the concept?

" 'I see' has always had two distinct meanings: to perceive with the eye and to understand with the mind" (Steen 1990, p. 2). The use of graphing calculators provides impressive opportunities for students to "see" with their eyes and to connect graphic images, symbolic expressions, and sets of related numerical values to compose mathematical pictures in their mind. What kind of mathematical comprehension will this rich imaging foster?

To demonstrate the fact that $\sin(\pi/2 - x) = \cos(x)$, students can compare the graph of $y = \sin(\pi/2 - x)$ to that of $y = \cos(x)$. If students can do this on their calculator, why teach them to prove identities? What is important about showing a sequence of steps to justify $(\tan^3 x + 1)/(\tan x + 1) = \sec^2 x - \tan x$? What is important about identities? Are there other ways to present those concepts? It would seem realistic to expect students to recognize that for all x, $\sin^2 x + \cos^2 x = 1$, but what other relationships are important enough to be learned for their own sake? Is it important for students to find $\cos \beta$ if $\tan \beta = \frac{3}{7}$ using the Pythagorean relationship to obtain $\frac{\pm 7\sqrt{58}}{58}$, when the inverse function key will produce a β that can be used as a reference angle? Is it important for students to solve quadratic equations by completing the square for equations such as $3x^2 - 5x - 9 = 0$? What mathematical concept will be enhanced by teaching these skills to students?

Using calculators, students should be able to view old ideas in new ways: If the limit of $\frac{(1 - \sqrt{x})}{(1 - x)}$ as x approaches 1 can be observed as $\frac{1}{2}$ from the graph of the function, why is it necessary to rationalize the numerator to obtain an equivalent expression and then find the limit by substitution? How important is finding limits algebraically rather than using graphs or zooming in on a table of values?

NEW POSSIBILITIES WITH CALCULATORS

Calculators also allow us to introduce mathematical ideas and techniques that have been beyond the reach of traditional low-tech curricula. For instance, parametric equations are not taught in most current algebra courses, yet they can be very helpful in graphing important relations that are not functions. Should parametric equations become a part of the curriculum or should students graph circles using $y1 = \sqrt{16 - x^2}$ and $y2 = \sqrt{16 - x^2}$? One possible answer can be found in the article by Foley in this volume.

Calculator operations are easily repeated, allowing students to investigate the behavior of numbers and functions recursively. Does this mean that the study of dynamic systems should become a part of geometry? Typically, complex numbers have had little relevance to any applications. Should they be introduced and linked to fractals? When and for what set of students?

Calculators can help with the analysis of sets of data entered as single variables, as ordered pairs, or as arrays. The data can be edited, sorted, and used in conjunction with calculations and graphing, which in essence allows real data from real problems to be used in the curriculum. If the random number function is used, calculators can also generate data, so simulations based on random numbers are easily accessible. How and where do these ideas fit into a curriculum that is presently tied to skill in manipulation and the formal development of mathematical systems?

Everyday experience is filled with sources of interesting data, and calculators can help produce histograms and other plots of these data. But since the calculator will not advise students about conclusions that can be drawn from the displays, where do students learn how to make judgments about the quality of the data or to interpret information displayed in a variety of ways?

Consider the following situation. Mike, a senior in a calculus class, was to analyze the set of data in table 3.1 and find an equation to summarize the data and the rate of growth for the population. Using his calculator, he plotted the data as ordered pairs, experimented with various models using regression calculations, graphed the resulting curves through a scatterplot of the data, and finally selected as the best model $y = (7.22613623\text{E}-5)\, 1.0090006425^x$. What mathematics did he use to make that decision? Is that mathematics currently a part of the curriculum in most schools? What does E − 5 mean? When can the decimals be rounded or truncated? Would $y =$

Table 3.1

Year	1800	1900	1920	1930	1940	1950	1960	1970	1980	1986
World population in millions	900	1625	1862	2070	2295	2513	3049	3704	4433	4917

(0.00007) 1^x give equally reliable results? These questions surface whenever students are given application problems and encouraged to use a calculator. How do teachers know what answers to give? What choices to make? How do they guide students to make valid decisions and interpretations about their numbers?

THE IMPACT ON TEACHING, LEARNING, AND TESTING

"Many mathematics educators worry that extensive use of calculator and computer tools, with corresponding de-emphasis of training in skills, will undermine development of conceptual understanding, proficiency in solving problems and ability to learn new advanced mathematics" (Fey 1990, p. 78). This concern is not easily addressed, yet it is apparent that a calculator can make significant contributions to the learning process. The graph of the function $y = \frac{(x - 1)}{(x^2 + 2x - 3)}$ would seem to make the distinction between an essential and a removable point of discontinuity much more readily than a symbolic analysis. The questions asked by students using calculators often reveal a perspective of mathematics much different from those raised in the traditional classroom. For example, one student who discovered the relation between slopes of perpendicular lines asked whether there were similar relationships between lines intersecting at any given angle. Encouraged by his teacher and using a calculator to explore his question, he found results that were exciting enough to be submitted to the local university for analysis. How often in the standard curriculum have typical high school students had opportunities for such rich and accessible investigations in mathematics?

An understanding of the errors a calculator can make in graphing a function (try graphing $y = \cos 76x$ in $[-6.28, 6.28]$) demands an understanding of the mathematical concepts involved. The search for the secant button, the graph of the greatest integer function, the appearance of asymptotes, and a zero result for an expression such as $10^{50} + 812 - 10^{50} + 10^{35} + 511 - 10^{35}$ offer insights into the nature of mathematics that were previously not available to students. Those who blindly push buttons without recognizing the mathematics in the situation will "soon learn that one of the fundamental principles of computers applies to calculators as well: Garbage in, garbage out" (Dion 1990, p. 567).

In a typical textbook chapter on irrational numbers, students spend much time applying rules to simplify irrational numerical expressions. They simplify $\sqrt{20}$ and $18/(\sqrt{3} - 5)$ and $\sqrt{2/3}$. Yet in reality most students have no understanding of irrational numbers. Consider the following experiment: When asked how many numbers were represented by $2 + \sqrt{3}$, 57 percent of the students in my school's twelfth-grade precalculus and calculus classes identified two numbers: 2 and 3. Another 21 percent wrote that there were three numbers: 2, 3, and $\sqrt{3}$. Only 21 percent explained that the sum created one irrational number. Perhaps it would be far better to have students approximate irrational numbers with calculators until the symbolic representation becomes meaningful.

One of the major issues raised by the prospect of calculator use in school mathematics is the impact of their use on standardized tests, most of which at present do not allow the use of a scientific calculator, let alone a graphing calculator. This issue is explored in more depth in the article by Greenes and Rigol in this volume.

Although some progress is being made, the situation is not likely to change in any major way for at least the next several years. Problems like the following occur on such tests today:

- Factor: $112a^3b + 26a^2b^2 + 10ab^3$
- Simplify: $\dfrac{\sqrt{x}}{(3\sqrt{x} - \sqrt{y})}$

The preview placement test for a large state university contains similar problems:

- Simplify: $(x - 2)(x^2 + 3) - 2x(x + 5)$

$$\frac{x - \dfrac{4}{x}}{\dfrac{x}{2} + 1}$$

If a teacher chooses to teach mathematics using a calculator, students may not be able to deal in a timely and efficient way with such problems. And it is important that students score well on those tests; for many students, admission to college or a chance for a scholarship hinges on their performance. Until changes occur in the tests and in the interpretations of the results of such tests, how can these different perspectives be resolved?

Calculator use in classroom assessment raises another important issue: How much should students record when using a calculator? What information will truly indicate to the teacher how the student has done the problem? What kind of a paper trail will allow the teacher to diagnose poor problem-solving strategies or misconceptions?

What does the calculator mean to a teacher? It should allow students to explore and conjecture. Students should be given opportunities to investigate situations, experiment with solutions, and discover mathematical relationships. Unless we teach our students how to deal with mathematics on calculators, however, they will not have any better understanding than they did with pages of manipulations. It has been claimed that factoring is a major topic in first-year algebra only because it can be taught. Are ideas now being taught because they can be taught using calculators? Will polynomial division have the same fate, as students wrestle with transforming

$$\frac{(x^2 + 1)}{(x - 1)} \text{ into } x + 1 + \frac{2}{(x - 1)}$$

in order to appreciate the graph of the expression? Will we force students to solve systems of linear equations in two unknowns by using the calculator and row-and-column manipulations, even when the system might be far easier to solve with pencil and paper? Or will we teach them to use the inverse matrix without talking about the structure that validates the process? Are we about to trade one set of magic rules for another?

CONCLUSION

If using calculators is to enhance the teaching and learning of mathematics and to open doors to new mathematics, the questions raised above must be answered. But as technology continues to change, new questions will surface—new ways to view old ideas and new conceptions of what is important will continue to emerge. Truly, teachers must be ready to learn along with students; they must accept questions that have no ready answers and be ready to deal with new ideas as well as malfunctioning calculators. Perhaps the only certainty is that as new directions for school mathematics unfold, Cheryl, Paul, Kim, and Adam and their teacher must learn to accept the calculator as a powerful tool they can employ to make learning and doing mathematics a rewarding and valuable experience. Teachers and students alike will find the kind of exciting surprises I encounter in calculator-enriched classes.

REFERENCES

Dion, Gloria. "The Graphics Calculator: A Tool for Critical Thinking." *Mathematics Teacher* 83 (October 1990): 564–71.

Fey, James T. "Quantity." In *On the Shoulders of Giants*, edited by Lynn Steen. Washington, D.C.: National Academy Press, 1990.

National Council of Teachers of Mathematics. *Curriculum and Evaluation Standards for School Mathematics*. Reston, Va.: The Council, 1989.

National Research Council, Mathematical Sciences Education Board. *Reshaping School Mathematics*. Washington, D.C.: National Academy Press, 1990.

Steen, Lynn, ed. *On the Shoulders of Giants*. Washington, D.C.: National Academy Press, 1990.

4

Research on Calculators in Mathematics Education

Ray Hembree
Donald J. Dessart

IN THE spring of 1984, we began to study all the research that could be found on the effects of calculator use in precollege mathematics. Our effort was prompted by what seemed to be conflicting circumstances: large potential benefits from the devices on the one hand; little actual impact on the curriculum on the other.

What had caused this lack of attention to calculators in the schools? As early as 1974, the National Council of Teachers of Mathematics (NCTM) had issued a statement that urged calculator use (p. 3), and ten years later, with shrinking costs, the device seemed ready to permeate American society. Negative forces appeared, however, that countered the positive attitudes. Mainly, these forces seemed to evolve from concerns that the handheld computing machine would displace students' skills with mental arithmetic and paper-and-pencil algorithms. The resulting debate about calculator effects induced a flurry of research, one of the largest efforts for any topic in mathematics education (Suydam 1982). Do calculators threaten basic skills? The answer consistently seemed to be no, provided those basics have been developed first in a calculator-free environment (Suydam 1979). Nonetheless, the question persisted. Some studies had yielded ambiguous findings. Other studies, although recording no harmful effects from calculators, failed to show improvement in either student achievement or attitude. Moreover, these findings were scattered and disparate in the literature, not centralized and integrated to tell a common story. What seemed to be needed was a rigorous, formal study of the overall body of research literature on calculator effects. Perhaps such an analysis would help resolve the controversy.

We found seventy-nine reports of experiments and relational investigations. In each of those studies, one group of students had been permitted to use calculators within a "treatment" period (in the usual case, about thirty school days), using the devices for computation or to help develop concepts and problem-solving strategies. In each study, during the same period, a comparison group received instruction on the same mathematical topics but had no in-class access to calculators. At the end of the treatment, both groups were examined; their average scores could then be compared.

The studies used a broad range of scales and instruments to measure outcomes. To synthesize the various findings, we transformed results to a common numerical base called *effect size*, with positive effect size indicating a study favoring the calculator treatment. In all, 524 "effects" were measured in the seventy-nine studies. These effects were then partitioned into subsets by grouping studies that focused on common aspects of performance or attitude. Within each subset, studies were clustered according to school grade and student ability levels. Average effect sizes were determined for each subset, and these effect-size averages were tested for statistical significance. The resulting average values were *the effects* of calculators in precollege classrooms. (See Hembree and Dessart [1986] for a formal report.)

Table 4.1 summarizes the findings. Performance had been measured in three aspects of mathematics: computation, concepts, and problem solving (as gathered, for example, from subtests of the Iowa Test of Basic Skills). In most of the studies, neither group was permitted to use calculators on tests. Thus, differences in their average scores revealed the effects of calculator use during instruction on students' skills with paper and pencil. In some of the studies, the experimental group was also allowed to use calculators during tests, whereas comparison students worked the same tests with only paper and pencil. The difference in average scores in those studies showed the advantage of using calculators on tests that followed instruction with calculators.

The seventy-nine studies showed no effects of calculator use on tests of

Table 4.1
Calculator Effects on Student Achievement (Original Study)

		Skills area	
Test condition	Student ability	Computation	Problem solving
With calculators*	Low	Moderate (+)	Moderate (+)
	Average	Large (+)	Small (+)
	High	No data	Moderate (+)
Without calculators	Low	Not significant	Not significant
	Average		
	Grade 4	Small (−)	Small (+)
	Other grades	Small (+)	Small (+)
	High	Not significant	Not significant

*Calculator groups only. Noncalculator groups used paper and pencil on tests.
Note: (+) and (−) denote respectively higher and lower scores of calculator groups compared to noncalculator groups.

conceptual knowledge. Effects for computation and problem solving were specific to student ability and school grade levels. For tests *with* calculators, scores were improved for low- and average-ability students, with effects that seemed moderate to large (3 to 8 points on a 100-point scale, converted from effect-size notation). None of the studies provided data on high-ability students in computation, but scores of these students in problem solving displayed a moderate improvement as a result of calculator use. Regarding tests *without* calculators:

1. For average students, small but significant effects (1 or 2 points on a 100-point scale) were observed at all grade levels. Three of the four effects were positive; the use of calculators during instruction advanced the students' skills with paper-and-pencil algorithms. The sole negative effect-size (based on a total of seven studies) regarded computation in grade 4. In that situation, calculators appeared to detract from the growth of students' computational skills.
2. No apparent effects were observed for low- or high-ability students in either computation or problem solving. The prior use of calculators had neither damaged nor improved the students' performance.

We interpreted these findings as encouragement toward calculator use in the classroom with a modest caution in some areas. On the positive side, the calculator could apparently advance the average student's computational skills while doing no harm to the computational skills of low- and high-ability students. Moreover, at all ability levels, calculators could provide a clear advantage when used during tests. The single negative finding serve to remind us that calculators, though generally beneficial, may not be appropriate for use at all times, in all places, and for all subject matters. Discretion in using calculators was advised.

Along with effects on performance, the body of studies had also yielded effects of calculator use on students' attitudes. Again, comparisons were drawn at the end of a treatment period after one student group had used calculators, whereas a second group had no such access. Results appear in table 4.2. Those students using calculators displayed a better attitude toward

Table 4.2
Calculator Effects on Student Attitudes (Original Study)

Dimension of attitude	Size of effect
Attitude toward mathematics	Small (+)
Self-concept in mathematics	Moderate (+)
Anxiety toward mathematics	Not significant

Note: (+) denotes a better attitude or feeling of calculator groups compared to noncalculator groups.

mathematics and an especially better self-concept in mathematics than students who had no formal contact with the devices.

EXTENDING THE STUDY

Subsequent to our meta-analysis, we have found nine additional studies that probed effects of calculator use in precollege mathematics (Bartos 1986; Bitter and Hatfield n.d.; Colefield 1986; Frick 1989; Heath 1987; Hersberger 1983; Magee 1986; Mellon 1985; Szetela and Super 1987). We used the results of these studies to extend our previous findings. Table 4.3 updates the outcomes related to student achievement. In every instance, new data either supported or enhanced the previous findings. For tests *with* calculators, the new data showed—

1. continued advantage from calculators in computation;
2. better advantage from the devices in problem solving.

For tests *without* calculators, the data suggested that using the calculator during instruction may improve paper-and-pencil skills for low- and high-ability students in addition to those of average ability. However, these findings were not strong enough to revise the prior interpretations.

Table 4.3
Updated Calculator Effects on Student Achievement

		Skills area	
Test condition	Student ability	Computation	Problem solving
With calculators*	Low	[Moderate (+)]**	[Moderate (+)]***
	Average	[Large (+)]**	[Small (+)]***
	High	No new data	[Moderate (+)]***
Without calculators	Low	[Not significant]**	[Not significant]**
	Average	—	—
	Grade 4	No new data	No new data
	Other grades	[Small (+)]**	[Small (+)]**
	High	[Not significant]**	[Not significant]**

*Calculator groups only. Noncalculator groups used paper and pencil on tests.

Notes: 1. Previous findings are shown in brackets. (+) and (−) denote respectively higher and lower scores of calculator groups compared to noncalculator groups.

2. ** denotes support of previous findings.
***denotes extension of previous findings in the same direction.

With regard to student attitudes, the data supported previous findings that calculators help promote a better attitude toward mathematics and an especially better self-concept in mathematics.

A SURVEY OF SURVEYS

Along with the meta-analysis of achievement and attitude data, the 1980s and early 1990s delivered descriptive findings on how well mathematics education has taken the calculator to heart. Results were usually gathered by surveys involving the following areas: (1) policies toward calculator use, (2) the accessibility of calculators in the schools, (3) modifications to the curriculum due to calculators, (4) frequencies of actual use by teachers in the classroom, (5) predominant activities with calculators, and (6) students' feelings toward the devices. This section summarizes such surveys. It should be noted that the older results may indicate trends more than current status.

Policy Statements

As the 1980s began, perhaps half the schools in the United States had declared a formal policy with regard to calculators (Jaji 1986). In contrast, more than 80 percent of Swedish schools had written policies. Only a few U.S. schools had expressly forbidden calculators, but Japan (land of the abacus) seemed very much opposed to the devices. As the decade continued, more and more schools viewed their use in a positive light. Few states have gone so far as to mandate their implementation, but at least 64 percent of the states have recommended their use for instruction in high schools, and 50 percent have suggested their use in all grades, K–12 (Kansky 1987). At least twelve states have also proposed that calculators be used in testing. For further description of the experience in Connecticut, Florida, and Michigan, see the respective articles by Leinwand, Hopkins, and Payne in this yearbook.

Calculator Availability

By 1987, only Connecticut had earmarked funds for the statewide purchase of calculators, and only six states had formally recommended calculator purchases (Kansky). The year before, research in Hawaiian classrooms showed little more than 20 percent equipped with calculators (Hawaii University Research Council). Recent surveys have indicated that access to calculators is probably less a problem than the earlier surveys suggested. Bitter and Hatfield (1991) reported a median value of more than three calculators available to a selection of seventh- and eighth-grade students, counting devices both at home and in school.

Curricular and Instructional Changes

By 1987, 42 percent of the states had produced guidelines or model curricula for aiding the integration of calculators into mathematics instruction (Kansky). Two states restricted calculator implementation to the upper grades (7–12), but the typical policy has recommended calculator use across the entire precollege spectrum, stressing the calculator as a *tool,* but not as an *object* of study. Maury's (1988) dissertation suggests that guidelines by the states do not seem well implemented at local school levels. In a selection of secondary schools of thirteen states, only 6 percent of the teachers recorded a fairly substantial impact (ten or more curricular changes) due to calculators. Eighty percent of the teachers reported five or fewer changes, and 40 percent reported zero impact on the curriculum. Bitter and Hatfield (1991) have also reported that although calculators seem to be prevalent, their regular use in classrooms is seldom the case.

Regarding change in instructional practice, twenty-two states have acted to (1) furnish information on new materials and techniques for teaching with calculators, (2) provide teacher in-service programs to study calculator technology, and (3) revise teacher certification standards to call for preparation in the use of calculators (Kansky 1987).

Frequencies of Actual Use

In 1980, NCTM urged classrooms at all grade levels to take advantage of calculators. Indeed, all grades have begun to use the devices. However, calculator use has not been uniform; higher grade levels apply calculators consistently more than the earlier grades. Teachers' willingness to teach mathematics with calculators has seemed to increase across all grades. In a 1980 study, percents of teachers who used calculators to teach mathematics were as follows: 14 percent in primary grades, 23 percent in intermediate grades, 42 percent in junior high school, and 62 percent in senior high school (Reys, Bestgen, Rybolt, and Wyatt). Similar patterns were noted by Scott in 1983 and Jaji in 1986. However, in 1989 Ford found 55 percent of grade 5 teachers using calculators at least occasionally (Ford 1989), and in 1988 Maury found 75 percent of secondary school teachers using calculators at least once a week (Maury 1988).

Activities with Calculators

In early grades, reported calculator activities have seemed confined to checking paper-and-pencil calculations, developing skill at estimation, and problem solving (Balka 1983; Ford 1989). Middle school classes include those calculator uses as well as such activities as assistance in teaching place value, aid in learning number concepts, games with calculators, special projects, and problems with large and realistic numbers (Jaji 1986). With regard to particular topics, the larger proportions of students seem prone to use the devices with fractions, decimals, computation, percent, and word problems. Smaller proportions apply calculators in dealing with geometry, measurement, probability, and statistics (Bitter and Hatfield 1991).

At senior high school levels, the calculator has come to be used as a tool in direct instruction. Areas of study included roots of higher-order polynomials, probability and statistics, decimal patterns, and high-order thinking. The calculator also seems readily permitted for homework and in testing (Jaji 1986; Maury 1988).

Student Attitudes toward Calculators

Along with frequencies of use of calculators in classrooms, students' attitudes toward the device have also changed over the past decade. In the 1981 Second International Mathematics Study (SIMS), only one-third of grade 8 students thought calculators made mathematics more fun and would offer them help in learning a wide range of topics (Jaji 1986). Those negative feelings may have been rooted in notions that seemed to prevail at the time—that using calculators is cheating (Reys 1980) and that regular use of calculators would cause students to lose their skills at mental computation (Hedren 1985). Of grade 12 students surveyed in the SIMS, some 83 percent thought calculators made the study of mathematics more fun; as students advanced in school, it seems, they became more inclined to welcome the use of calculators. Nonetheless, they remained noncommittal whether or not calculators could help them learn a great many topics.

Attitudes toward calculators seem to have become more positive at the start of the 1990s. Table 4.4 shows results of a recent survey of nearly 500 middle school students (Bitter and Hatfield 1991). In large proportions, the students favored the presence of calculators during most mathematics activities. However, they seemed to display more reserve regarding calculators on tests and regarding questions on whether or not the calculator will hinder the understanding of basic computational skills. Most students believed that calculators ought to be used in restricted conditions, to check arithmetic computations, and to learn particular topics.

Table 4.4

Percent of Students Responding "Agree" or "Strongly Agree" to Attitude Items (from Bitter and Hatfield 1991)

Attitude item	Percent
Calculators make mathematics fun.	79.1
Mathematics is easier if a calculator is used to solve problems.	86.3
It is important that everyone learn how to use a calculator.	85.6
I would do better in math if I could use a calculator.	72.7
I prefer working word problems with a calculator.	69.6
I would try harder in math if I had a calculator to use.	49.3
Students should not be allowed to use a calculator while taking math tests.	28.3
The calculator will hinder students' understanding of the basic computation skills.	36.6
Since I will have a calculator, I do not need to learn to do computation on paper.	12.7

A NOTE ON NEW TECHNOLOGY

Manufacturers of calculators continue to produce machines whose special capabilities may enhance mathematics teaching and learning. An example of recent technology is the graphing calculator, which appears potentially useful in the study of high school algebra and beyond. Evidence of the reality in that promise is only beginning to emerge in the research literature, but it is a topic of urgent need. In one such study, seventy-three students in Utah used the graphing calculator in advanced algebra during their first semester of study (late 1990 and early 1991). One hundred eighty-four of their peers studied the same course material without using such devices. Test scores with paper and pencil at the end of the semester showed a 19 percent advantage for the group equipped with calculators during instruction (data provided by Pam Giles, Brighton High School, Salt Lake City). The study was observational in nature and used no statistical tests of significance. Nonetheless, it stands as evidence that the value of calculators in classrooms seems likely to be expanded through continued innovation (see also Browning [1990]; Rich [1990]).

CONCLUSIONS

We conclude this chapter with the following observations:

1. The preponderance of research evidence supports the fact that calculator use for instruction and testing enhances learning and the performance of arithmetical concepts and skills, problem solving, and attitudes of students. Further research should dwell on the best ways to implement and integrate the calculator into the mathematics curriculum.

2. In some of the synthesized research, calculators were used for instruction and then their use was denied for testing. To deny the use of calculators for both instruction and testing is an antiquated policy that should be changed.

3. The use of calculators in the early grades is frequently for familiarization, for checking work, and for problem solving. The senior high school seems to emphasize using calculators as tools for calculation and reference. There have been no empirical studies on how to integrate the calculator directly into the learning process. The favorable Salt Lake City findings on the use of graphing calculators bodes well for their future in algebra, but little other research has been reported on graphing calculators.

4. The attitudes of students using calculators are favorable, but apparently some students still feel that using a calculator is tantamount to "cheating." The latter is a most unfortunate attitude for students to develop.

5. Most schools possessing calculators tend to have a single classroom set of the devices. It seems clear that for most efficient use, a calculator should be made available for each student.

6. The years of the 1980s saw growth in the use of calculators in schools. It seems clear that this trend will accelerate through the 1990s.

Calculators are gaining wide acceptance by teachers. The findings of research probably have accelerated that acceptance. Further research should dwell on the uses of calculators for developing conceptual understandings, number sense, graphical sense, and estimation. The integration of the calculator into the curriculum where it plays a central role in the learning process is a worthy goal for the research of the 1990s.

REFERENCES

Balka, Don S. *The Status of School Mathematics.* Indianapolis: Indiana State Department of Public Instruction, 1983. (ERIC Document Reproduction Service No. ED 236 048).

Bartos, Joyce J. "Mathematics Achievement and the Use of Calculators for Middle Elementary Grade Children." *Dissertation Abstracts International* 47(1986): 1227A.

Bitter, Gary G., and Mary M. Hatfield. "The Calculator Project: Assessing School-wide Impact of Calculator Integration on Mathematics Achievement and Attitude." Unpublished manuscript.

Browning, Christine C. "Characterizing Levels of Understanding for Functions and Their Graphs." Doctoral dissertation, Ohio State University, 1990.

Colefield, Ronald P. "The Effect of the Use of Electronic Calculators versus Hand Computation on Achievement in Computational Skills and Achievement in the Problem-solving Abilities of Remedial Middle School Students in Selected Business Mathematics Topics." *Dissertation Abstracts International* 46(1986): 2168A.

Ford, Margaret I. "Fifth-Grade Teachers and Their Students: An Analysis of Beliefs about Mathematical Problem-solving." *Dissertation Abstracts International* 50(1989): 378A.

Frick, Faye A. "A Study of the Effect of the Utilization of Calculators and a Mathematics Curriculum Stressing Problem-solving Techniques on Student Learning." *Dissertation Abstracts International* 49(1989): 2104A.

Hawaii University Research Council. *Mathematics Instruction in Tokyo's and Hawaii's Junior High Schools. Final report.* Honolulu, Hawaii: University of Hawaii & National Institute for Educational Research, Tokyo, Japan. (ERIC Document Reproduction Service No. ED 269 251), 1986.

Heath, Robert D. "The Effects of Calculators and Computers on Problem-solving Ability, Computational Ability and Attitude toward Mathematics." *Dissertation Abstracts International* 48(1987): 1102A.

Hedren, Rolf. "The Hand-held Calculator at the Intermediate Level." *Educational Studies in Mathematics* 16(1985): 163–79.

Hembree, Ray, and Donald J. Dessart. "Effects of Hand-held Calculators in Precollege Mathematics Education: A Meta-Analysis." *Journal for Research in Mathematics Education* 17 (1986): 83–99.

Hersberger, James R. "The Effects of a Problem-solving Oriented Mathematics Program on Gifted Fifth-Grade Students." *Dissertation Abstracts International* 44(1983): 1715A.

Jaji, Gail. *The Use of Calculators and Computers in Mathematics Classes in Twenty Countries: A Source Document.* Second International Mathematics Study. Urbana, Ill., Illinois University, and Washington, D.C.: Center for Educational Statistics, 1986. (ERIC Document Reproduction Service No. ED 291 590).

Kansky, Bob. *Technology Policy Survey: A Study of State Policies Supporting the Use of Calculators and Computers in the Study of Precollege Mathematics.* Reston, Va.: National Council of Teachers of Mathematics, 1987. (ERIC Document Reproduction Service No. ED 289 728).

Magee, Elaine F. "The Use of the Minicalculator as an Instructional Tool in Applying Consumer Mathematics Concepts." *Dissertation Abstracts International* 47(1986): 455A.

Maury, Kathleen A. "The Development of an Instrument to Measure the Impact of Calculators and Computers on the Secondary School Mathematics Curriculum." *Dissertation Abstracts International* 49(1988): 755A.

Mellon, Joan A. "Calculator-based Units in Decimals and Percents for Seventh Grade Students." *Dissertation Abstracts International* 46(1985): 640A.

National Council of Teachers of Mathematics. *An Agenda for Action: Recommendations for School Mathematics of the 1980s.* Reston, Va.: The Council, 1980.

———. "NCTM Board Approves Policy Statement on the Use of Minicalculators in the Mathematics Classroom." *NCTM Newsletter* 11 (December 1974): 3.

Reys, Robert E. "Calculators in the Elementary Classroom: How Can We Go Wrong!" *Arithmetic Teacher* 28 (November 1980): 38–40.

Reys, Robert E.: Barbara J. Bestgen, James F. Rybolt, and J. Wendell Wyatt. "Hand Calculators: What's Happening in Schools Today?" *Arithmetic Teacher* 27 (February 1980): 38–43.

Rich, Beverly S. "The Effect of Using Graphing Calculators on the Learning of Function Concepts." Doctoral dissertation, University of Iowa, 1990.

Scott, Patrick B. "A Survey of Perceived Use of Mathemátical Materials by Elementary Teachers in a Large Urban School District." *School Science and Mathematics* 83(1983): 61–68.

Suydam, Marilyn N. *The Use of Calculators in Precollege Education: A State-of-the-Art Review.* Columbus, Ohio: Calculator Information Center, 1979. (ERIC Document Reproduction Service No. ED 171 573).

———. *The Use of Calculators in Precollege Education: Fifth State-of-the-Art Review.* Columbus, Ohio: Calculator Information Center, 1982. (ERIC Document Reproduction Service No. ED 220 273).

Szetela, Walter, and Doug Super. "Calculators and Instruction in Problem-solving in Grade 7." *Journal for Research in Mathematics Education* 18(1987): 215–29.

Part 2

5

CAN: Calculator Use in the Primary Grades in England and Wales

Hilary Shuard

In England, impetus for developing a calculator-aware number (CAN) curriculum for children in the primary years came from the Cockcroft Report on the Teaching of Mathematics in Schools in England and Wales (Department of Education and Science 1982). When this important national report was published, calculators were becoming everyday tools for adults but were not yet being widely used in schools. Among many other issues in mathematics teaching, the Cockcroft Report discussed the use of calculators in primary schools and recommended that development work should be undertaken:

> There is as yet little evidence about the extent to which a calculator should be used instead of pencil and paper for purposes of calculation in the primary years; nor is there evidence about the eventual balance to be obtained at the primary stage between calculations carried out mentally, on paper, or with a calculator. However, it is clear that the arithmetical aspects of the primary curriculum cannot but be affected by the increasing availability of calculators. (Para 387)

In the next few years it became clear that the arithmetical aspects of the primary curriculum were not yet being affected by calculators, which were still not being used in most primary schools. The recommendations of the Cockcroft Committee about the use of calculators had largely been ignored. An opportunity for development work came when the PrIME Project was set up. PrIME was a nationally funded curriculum development project in primary school mathematics, which operated in England and Wales from 1985 to 1989. One of its aims was

> to develop the primary mathematics curriculum to take full account of the impact of new technology, concentrating especially on the importance of calculators for the number curriculum.

In pursuit of this aim, the project set up working groups of teachers to investigate the effect of a long-term complete acceptance of calculators in the primary school classroom, from the age of six. It was hoped that the project would be able to continue until the first children entered secondary schools at the age of eleven. In fact, the development of CAN will continue until the first children are aged at least twelve.

Initially, in September 1986, the project involved twenty classes of six-year-old children and their teachers. Schools were asked to ensure that these classes would be able to remain in CAN until the children went to secondary schools, by which time they would have worked at CAN for five years. In each school, all successive groups of six-year-olds have joined the project, by the school's own choice. Thus, CAN is subject to all the usual changes and chances of school life and has been taught by the usual variety of teachers found in primary schools. However, schools that volunteer for such a project are normally above average in enterprise and enthusiasm. The schools are grouped in clusters of three or four in different parts of the country, and they span the full range of social conditions found in Britain.

At the beginning of the project, each child was provided with a simple four-function calculator, and teachers were asked always to allow the children to decide for themselves whether to use their calculators or to calculate in other ways. The teachers were also asked not to teach the traditional pencil-and-paper vertical algorithms for addition, subtraction, multiplication, and division; it was clear that children would no longer need these algorithms, since calculators would always be available. The large amount of time previously taken up with the practice of algorithms was released, and this enabled the teachers to work on developing children's understanding of number and to give them opportunities for the exploration and investigation of mathematics.

In 1986, there was very little experience, anywhere in the world, of calculator use by very young children, so it was difficult to predict what would happen in CAN. However, project teachers were urged to continue with "good primary practice," which included providing a large range of hands-on, practical number activities, encouraging the investigation of mathematical problems, and developing children's mathematical language through discussion. In CAN, children would also be encouraged to develop their powers of mental calculation and to share their methods with one another.

If a published textbook scheme was in use in the school, teachers were asked not to use it for number work, since the number work in all existing textbook schemes was based on teaching the traditional pencil-and-paper algorithms; these schemes were therefore unsuitable for CAN. However, teachers were not discouraged from using published schemes for other aspects of mathematics, such as measurement, graphical work, and shape and space. The CAN mathematics curriculum was intended to be a broad one, so that number work would be not more than half of it.

THE TEACHER'S ROLE

The CAN project team believed that the most suitable people to undertake curriculum development were the teachers who worked with the children every day in the classroom and who could respond immediately to the children's ideas and needs. Consequently, cooperative groups of teachers were set up in each area. Each group worked together to develop both their teaching and their curriculum. A style of working emerged that depended on mutual support and the sharing of ideas within a group of teachers. The central project team has not furnished classroom materials for CAN, but it has tried to ensure that teachers were supported in developing and sharing their own ideas and their own materials.

The teachers' groups meet regularly, at about monthly intervals, to exchange ideas and discuss ways of tackling classroom problems. Most clusters have an "advisory teacher" attached to them, who visits classrooms and supports the teachers in their work with children. The project central team has attended teachers' meetings and visited CAN classrooms whenever possible in order to share ideas among the groups. However, most of the development of CAN has been undertaken by classroom teachers; they have found the regular meetings with colleagues an essential support. At first, project teachers devised fairly formal activities to introduce children to the calculator. However, the teachers soon began to see the possibility of using open-ended "starting points" and to appreciate the opportunities these gave children. With open-ended starting points, children could use numbers of their own choice and could develop the work in their own way. Activities and starting points were shared at teachers' meetings and often gave rise to further ideas.

Some themes were frequently discussed at teachers' meetings. These included how to respond to children who ask difficult questions about such topics as negative numbers and decimals, children's methods of recording their work, and teachers' surprise at what their children could do. One teacher of six-year-olds wrote about this experience:

> Prior to CAN I tended to limit my teaching to numbers up to 20. I was therefore surprised to find a group of children counting confidently in hundreds when weighing objects using 100-gram weights.

Since the teachers were released from the need to teach the traditional pencil-and-paper algorithms, they had time to introduce much more problem-solving and investigative work and to develop their own teaching styles. The teachers became more responsive to children's ideas and began to see their role as participants in the children's work rather than as instructors who told children how to do mathematics. They developed a style of talking with children about mathematics that was quite different from the usual question-answer-evaluation style of classroom discourse. The teachers no longer pointed children toward the expected "correct answers" to the teacher's questions but instead asked the children to explain their own thinking. The teachers became expert in responding to children's questions in such a way that the children were valued and supported but required to do the mathematical thinking for themselves. As this style developed, the teachers became confident in their own thinking. One teacher wrote:

> Through developing activities, considering children's responses, and placing a greater emphasis on listening and talking to children and trying to understand how they think mathematically, I feel more confident in teaching mathematics.

SOME TRENDS IN THE MATHEMATICS CURRICULUM

Abandoning the traditional arithmetical algorithms has had an exciting effect on the mathematics curriculum and the children's confidence. Children who use calculators do not get wrong answers (except through errors of problem conceptualization or calculator keystroking) or worry about the difficulties of calculation. The children have learned that in mathematics you need to think for yourself rather than imitate the teacher. The calculator has provided an exploratory introduction to topics that are not part of the traditional curriculum for young children but that a calculator user cannot avoid encountering. These topics include large numbers, negative numbers, decimals, and square roots. The video produced by the project (Shuard et al. 1991b) shows an episode in which a teacher talked to his class of eight-year-olds:

> *Teacher:* Who thinks they know the biggest number in the world?
>
> *Child:* There isn't one.
>
> *Teacher:* Why not?
>
> *Child:* You can always add 1 to any number you've got.

Many children have become fascinated with number patterns, which they can explore easily, using the calculator when necessary. Another trend has been the interest that many children show in mental calculation and other methods of noncalculator calculation. They want to be able to "do it themselves" rather than rely on the calculator. Because they have not been taught the traditional algorithms, they have relied on their own ingenuity and understanding of numbers; many of their methods are sound and elegant. Examples are given in the next section.

Another effect of CAN has been the growth in problem-solving and investigative work and in the use of mathematics in cross-curricular topics. For example, one group of eight-to-nine-year-olds tackled the following investigation:

> Make numbers by the addition of consecutive numbers, like this:
>
> $$3 + 4 + 5 = 12 \quad \text{and} \quad 6 + 7 = 13$$
>
> Investigate.

Most children found that they could not make 2, 4, 8, and 16, and they predicted that 32 and 64 would also be impossible. Rebecca attempted to reason logically about the situation. She wrote:

> The ones so far I can't make are 4, 8, 16, and they are all even. I think that the reason that we can make all the odd numbers is because that when you take an even number and an odd number and you add them together you always get an odd number. Though when you take two odd numbers they always make an even number but we are not allowed to take odd numbers together because they are not next to each other.

CHILDREN'S WORK

In this section, examples are given of the written work of children aged six to nine, grouped by topics. The topics are chosen to show aspects of mathematics that do not feature prominently in the traditional primary curriculum. It is important to realize that written work is only one aspect of the children's mathematics—they also do a great deal of practical work, using a variety of apparatus.

Place Value and Large Numbers

A popular activity for young CAN children is the following:

> Put a number inside a square. Then put numbers at each corner of the square so that those numbers add up to the number in the square.

Although this is not a calculator activity, it can be checked on the calculator. Gary's way of doing it (fig. 5.1) was quite a surprise to his teacher, who had not yet "done" any work with these six-year-olds on place value in hundreds. However, Gary seemed to have found out how to decompose a three-digit number into hundreds, tens, and units, although he was not yet sure how to write a "7."

Fig. 5.1

At the same time, in the same class, Sara was experimenting with seven- and eight-digit numbers and devising her own variation on the activity (fig. 5.2).

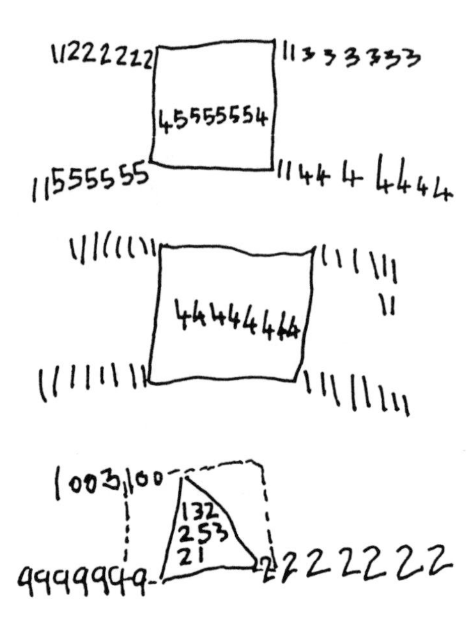

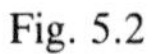
Fig. 5.2

A teacher wrote about a six-year-old's work on an activity about the amount of water drunk at lunchtime:

> I chatted to each group and asked them to explain how they had worked it out. One girl seemed to be in difficulties. She had written on her paper
>
> 10 glasses was 1200 millilitres
>
> so I asked her how much 20 glasses would be. It was quite a time before she said, "Two of those," and wrote
>
> 20 glasses was 2400 millilitres.
>
> I asked how much 30 glasses would be. After what seemed like eternity, when she was working on another task, she wrote
>
> 30 was 3600 ml.
>
> I was really pleased when she read out the number as "3 hundred and 60 ml" and said, "No, it isn't, it's 3 thousand 6 hundred."

For many children, the correct way of writing numbers in the thousands is not at first obvious. Peter started with a four-spike abacus and arranged two beads as shown in figure 5.3. He said, "It's one thousand one hundred. How do you write it down?" He continued, "I know; I'll use the calculator."

Fig. 5.3

He keyed in 1000 [+] 100 [=], and the display showed the information he needed.

Most children enjoy using the largest numbers they can handle confidently. Many CAN "starting points" encourage children to use numbers of their own choice, allowing the more able to extend their command of numbers while the less able remain within their own limits. Both the children whose work is shown in figure 5.4 were aged seven, and both were regarded as "slow learners." They were "making 19." Steve was much more confident than Stewart, but eventually the numbers grew beyond his control. Patterns like these are not usually generated on the calculator; indeed, the mistakes in the second half of Steve's work suggest that he did not even verify his pattern on the calculator.

Stewart	Steve
40 − 21 = 19	100 − 81 = 19
	1000 − 981 = 19
10 + 9 = 19	2000 − 1981 = 19
	3000 − 2981 = 19
20 − 1 = 19	4000 − 3981 = 19
	5000 − 4981 = 19
50 − 31 = 19	6000 − 5981 = 19
	7000 − 6981 = 19
20 − 1 = 19	8000 − 7981 = 19
	9000 − 8981 = 19
30 − 11 = 19	10000 − 9881 = 19
	11000 − 1081 = 19
13 + 6 = 19	12000 − 2081 = 19
	13000 − 3081 = 19
15 + 4 = 19	14000 − 4081 = 19
	15000 − 5081 = 19
17 + 2 = 19	16000 = 6081 = 19
	17000 = 7081 = 19
19 + 0 = 19	18000 = 8081 = 19
	19000 = 9081 = 19

Fig. 5.4

Another teacher told of a different aspect of her students' understanding of very large numbers:

> The children were talking about the biggest number they could get. Shaun had 10000000. The biggest numbers they could produce all began with 1. Thinking about the calculator keys, the teacher discussed "bigger than" up to 9. The children then realised that all 9s would give them a bigger number. Shaun, however, said, "I think the biggest number is billions and trillions, isn't it, Sally?" Sally (aged six) replied that she didn't think there was an end to numbers.

Negative Numbers

The CAN children were furnished with simple four-function calculators. The model used at first did not have a change sign (+/−) key, so that it was not possible to key negative numbers directly into the calculator. However, the children were not long in finding negative numbers. For example, the calculation 6 − 8 = gives the following display:

− 2

The children quickly spotted and asked about the negative sign, and number lines had to be extended below zero.

Soon, teachers were enquiring into children's knowledge of negative numbers. One teacher wrote:

> The children were subtracting from 50 using dice and Dienes blocks. They were trying to get to 0. They wrote down 50 and subtracted the number shown on the die, using Dienes blocks as a check on their mental calculation. Jenny had 3 left and shook 5. She said, "I can't take it away. I would owe 2." She tried this on a calculator and said, "It is take away 2." She later tried to make other negative numbers, and she could do this. When given the problem
>
> The answer is −1. What is the question?
>
> she produced a pattern:
>
> 1 − 2
> 2 − 3
> 3 − 4
> .
> .
> .
>
> When asked what needed to be taken away from 100 to give −1, she said: "Easy—101." She said she always made the second number one bigger. She could use this method when the answer was −2, but not for −3.

One teacher invented a game in which negative numbers were likely to turn up naturally. The game was called Smarties (Smarties are popular, brightly colored sweets); the rules were as follows:

Before starting this game, children count and sort a tube of Smarties, and make a block graph of the colors. Suppose there are 31 Smarties in the tube. Each child receives 31 points to start the game. The Smarties are returned to the tube at the beginning of the game. They are taken out one by one, and each child guesses which color will come out next and records it. Correct guesses score 2 points; incorrect guesses lose 2 points.

Children are encouraged to consult the graph and work out a probable color to appear next. They often record when all the Smarties of one color are out. Negative numbers appear naturally, and much of the calculation of scores is mental, but calculators and number lines are available for checking.

In one classroom, a problem arose when a child wanted to know what happened when two negative numbers, -3 and -4, were multiplied together. The teacher was unsure what the answer should be, and the basic four-function calculators available could not cope with the calculation. When a calculator that handled negative numbers was found, teacher and child learned together. Many children were very proud of their ability in mental calculation. Some began to make use of negative numbers in these calculations. The method of subtraction shown in figure 5.5 was discovered independently in several classrooms by children who prefer to work from the left-hand side of the numbers rather than the right-hand side. Many people, both adults and children, choose to start mental calculations from this end.

333 − 165 = 168
300 − 100 = 200
30 − 60 = −30
3 − 5 = −2
200 − 30 = 170
170 − 2 = 168

Fig. 5.5

Noncalculator Methods of Calculation

As the children have become familiar with numbers and with the operations of addition, subtraction, multiplication, and division, they have developed mental methods for these operations. They find the calculator unnecessary and slow when the numbers are within the range with which they can cope, and they are proud of their ability to "do it themselves." Of course, the size of the numbers they can handle mentally varies from child to child.

Although some children's mental methods of addition are rather like the traditional pencil-and-paper method of addition, no child has devised a method of subtraction that resembles the traditional subtraction method. Some examples of children's methods of calculation are given here. At the end of the first year of CAN, one teacher individually asked twelve seven-year-olds to work out 29 + 28 in their heads and explain how they had done it. Of these twelve children, two were unable to do the calculation in their heads, and the remaining ten children all had slightly different mental methods. For example:

- 30 and 30 are 60. Then you take away. Take away 1 for the 29, and take away 2 for the 28. So take away 3 altogether, and you get 57.
- 8 and 9 are 17. 20 and 20 are 40. 40 and 17 are 57.
- The 2 twenties are 4 tens, so that's 40. Add 9, which gives 49. Add 1, which makes 50. Then add the other 7, so that's 57.
- Two 25's are 50. Then you have to add on 3 and 4, so you have to add on 7. It's 57.

Of these four children, three started (in various ways) with the tens.

Without a calculator, multiplication and division of large numbers are much more difficult than addition and subtraction, and children may need to record intermediate results. However, some children have mental methods for multiplying by single-digit numbers, as two teachers explain. All the children they quote multiplied the tens before the units.

- I asked Errol to try 36 multiplied by 3 mentally. He quickly said it was 108, and explained that three 30s made 90, three 6s made 18, and 90 and 18 was 108. I then asked him to do 13 multiplied by 36. Taking a little longer, he said 468. He explained that it was four lots of 108 add 36.
- I asked the children to explain how they worked out 48 × 5. One said, "50 × 5 is 250, and you take away 10, so it's 240." Another child did 36 × 5 as 30 × 5 = 150 and then 6 × 5 = 30.

In the third year of CAN, a nine-year-old, Hannah, decided to develop a method of multiplying two 2-digit numbers without a calculator. She was proficient at adding and subtracting 2-digit numbers mentally, so she tried the same method. She set herself to multiply 54 by 62, and after she had completed the task, she wrote about what she had done (fig. 5.6). This account omits her learning process, but her teacher described what happened. She first multiplied 50 × 60 and obtained 3000; then she multiplied 4 × 2 and obtained 8. She added the two results together to give 3008. She then thought she had finished and checked her result on the calculator. She was extremely surprised that the calculator gave 3348. She puzzled about the difference of 340 between her result and the calculator; eventually she realized that the calculator had done two additional multiplications: 4 × 60

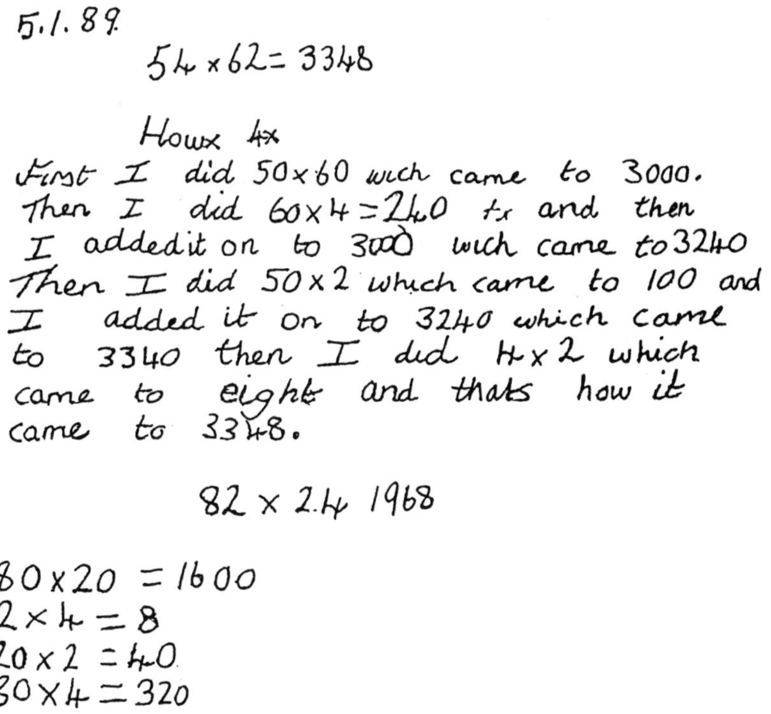

5.1.89

54 × 62 = 3348

How × 4×
First I did 50 × 60 wich came to 3000.
Then I did 60 × 4 = 240 ×× and then
I addedit on to 3000 wich came to 3240
Then I did 50 × 2 which came to 100 and
I added it on to 3240 which came
to 3340 then I did 4 × 2 which
came to eight and thats how it
came to 3348.

82 × 24 1968

80 × 20 = 1600
2 × 4 = 8
20 × 2 = 40
80 × 4 = 320

Fig. 5.6

and 50 × 2. Her own account gives the multiplications in a different order and applies the method to 82 × 24. Like other teachers, Hannah's teacher valued and encouraged the children's understanding of number and their ability to calculate by their own methods.

A QUANTITATIVE EVALUATION

The evaluation of the project has been qualitative; the evaluator observed and reported what she saw teachers and children doing, but she did not test groups of children. However, outside circumstances in one Local Education Authority (LEA) made quantitative evaluation inescapable. For some years in that LEA, all children had been tested in mathematics during the school year in which they attained the age of nine. Thus, the mathematical performance of the CAN project children on the 8+ Maths Test could be compared with the performance of other children of the same age.

The test was revised so that it would not give undue advantage to children who either did or did not habitually use calculators. It became a test of understanding not only of number but of a variety of mathematical ideas. CAN children first took the test in 1989, when they had been in the project for either one or two years. A total of 116 project children were tested, and their performance was compared with that of 116 other children chosen at random from the other children in the LEA.

In twenty-eight of the thirty-six test items, more project children than other children gave correct responses. In eleven items, the success rate of the project children was 10 percent or more higher than that of the other children; in one item it was 30 percent higher. In the remaining eight items, the nonproject children performed as well or better than the project children. However, the maximum percentage by which the other children exceeded the project children was 5.3 percent. In twenty items, a higher percentage of nonproject children did not attempt the item, suggesting that CAN project children are more able to work independently.

It is quite possible that the project children who took the 8+ Maths Test were not totally comparable with the children with whom they were compared. Thus, it would not be advisable to put too much weight on the result of this one test. However, the qualitative evaluation was also very positive. The evaluator commented particularly on the children's enthusiasm, their understanding of topics usually thought too difficult for their age range, their ability in mental calculation, and their persistence in tackling problems. For lower-attaining children, the evaluator commented that although CAN has not turned them into great mathematicians, it has greatly improved their attitudes.

PARENTS AND OTHERS

At the beginning of CAN, schools were asked to ensure that parents knew about the project. Most parents have been very supportive, although a few feel that life is impossible without pencil-and-paper algorithms; they may try to teach these at home—often to children who do not share their view! As the project has gone on, parents have become more supportive. The children have converted their parents; after a few months, parents started to say, "I was very doubtful at first, but now I can see what my child can do."

A problem that has troubled some schools has been the number of visitors who wish to see CAN in action. One teacher said to a child at the end of one day, "What have you done today?" The child replied, "I've talked to seventeen visitors." It became clear that in the children's interest, visiting would have to be limited.

CONCLUSION

A short article like this one leaves no room to describe such topics as the children's exploration of decimals, which they often describe as "the way the calculator writes fractions." Nor can the way the project is developing as the children get older be described. A fuller account of the project can be found in Shuard et al. (1991a).

In 1989 a national curriculum was introduced into English schools. This requires the use of three methods of calculation: mentally, with a calculator, and with pencil and paper. Although it is encouraging that all children will now be required to use calculators, it is disappointing that they will still be required to

> (using whole numbers) understand and use non-calculator methods by which a 3-digit number is multiplied by a 2-digit number and a 3-digit number is divided by a 2-digit number. (Department of Education and Science and Welsh Office 1989)

The pace of change in education remains slow, and it continues to be difficult to ensure that new thinking is taken into classroom practice. But the last word belongs to the child who said, "CAN maths is great. Much better than sums."

REFERENCES

Department of Education and Science. *Mathematics Counts: Report of the Committee of Inquiry into the Teaching of Mathematics* [The Cockcroft Report]. London: Her Majesty's Stationery Office, 1982.

Department of Education and Science and Welsh Office. *Mathematics in the National Curriculum*. London: Her Majesty's Stationery Office, 1989.

Shuard, Hilary, Angela Walsh, Jeffrey Goodwin, and Valerie Worcester. *Calculators, Children and Mathematics*. London: Simon & Schuster, 1991a.

———. *Calculators, Children and Mathematics*. Videotape. London: Simon & Schuster, 1991b.

6

A Constructivist Approach to Developing Early Calculating Abilities

Ian Sugarman

FOR many decades, the teaching of arithmetic algorithms has enjoyed an unchallenged place in the mathematics curriculum of British primary schools. Generations of schoolchildren have been expected to learn and display their ability to carry out calculations in very narrow and rigidly defined ways, not allowing for any individual creativity. However, the continued existence of that traditional arithmetic curriculum now seems seriously threatened.

Some attribute the possible demise of pencil-and-paper arithmetic algorithms to the arrival of inexpensive hand-held calculators. I prefer to view calculators as only the final nail in the coffin of those algorithms. More important is that teachers of primary schoolchildren are devoting their efforts to the development of mental rather than written skills of calculation.

The National Curriculum in the United Kingdom (Department of Education and Science 1989) gives official sanction to this new emphasis. The curriculum includes no statements of attainment for any specific pencil-and-paper algorithms, a position that is amplified by National Curriculum Council's (1989) *Non-Statutory Guidance:*

> In order to progress through the levels, pupils at every stage should be encouraged and helped to develop *their own* methods for doing calculations. (E1.4)

> This central place of mental methods should be reflected in an approach that encourages pupils to look to these methods as a *"first resort"* when a calculation is needed. Such methods are the basis upon which all standard and non-standard written methods are built, and they underpin a wide range of approaches to calculating. (E2.2)

This position is based, in part, on concern about pupils' lack of understanding of the symbol manipulations undertaken in traditional routines. Some teachers regret this state of affairs, drawing comfort from the *belief* that even if their efforts to secure understanding have been largely fruitless, at least teaching algorithms has equipped pupils with useful skills.

The fact that "transmission"-based teaching strategies depend primarily on pupils' powers of memory rather than on their creative powers of reasoning and judgment has led many mathematics educationists to look for an alternative teaching strategy in which the objective is the production of awarenesses as opposed to mere behavioral changes. My own work has sought to discover ways to help children construct their own strategies of calculation rather than to demonstrate ways in which specific arithmetical routines may be more effectively taught. In the realm of single- and two-digit calculation, it is taken that those strategies will essentially be mental rather than written and idiosyncratic rather than standardized.

A constructivist approach to developing children's mental strategies of calculation recognizes that students learn by a process of refinement and reconstruction. This process does not depend as much on direct teaching as it does on a more protracted process in which experiences are "made sense of" by referring to existing conceptual structures. Those structures, in turn, may be challenged and restructured to accommodate the new experiences.

This constructivist learning process can be accelerated in the classroom in at least three ways:

1. By providing opportunities to manipulate representations of number within a decision-making context
2. By a curriculum that emphasizes the detection of patterns and relationships
3. By encouraging pupils to express their perceptions of patterns and to share their personal strategies of calculation

Significantly, these three elements—decision making, prediction, and communication—are at the heart of the new National Curriculum for Mathematics in the British state education system.

MONITORING CHILDREN'S DEVELOPMENT

Exploring the territory of children's mental calculations has revealed a rich diversity of strategies (see Ginsburg [1982] and Ashfield [1989]). My own research has exposed a range of pupil misconceptions possibly derived from rote-based teaching strategies. These usually resulted in failure. All, however, appeared to involve the creative integration of a range of awarenesses, skills, and recalled facts currently at the pupil's disposal. This important information about individual children's understanding of number and strategies of calculation was obtained in the classroom as pupils worked on mathematical games activities. The calculator played a key role.

Adults and children alike are unaccustomed to reflecting on their mental processes, and pupils do not always find it easy to become aware of their own mental strategies of calculation. Experience has shown, however, that such reflective skills improve with practice, and children who are regularly asked by their teacher to describe their calculating strategies soon learn to do so routinely.

In the assessment activity described in the next section, the teacher asks a pupil to describe how she came to choose the operation she is about to enter (or has just entered) into the calculator. The child then attempts to reconstruct her thinking while the teacher makes a mental or written note.

This method of investigation is in sharp contrast to the practice of some researchers whose primary concern has been to discover the efficacy of a particular teaching program based on instruction rather than pupils' ability to calculate per se. In that type of research, the form in which the problems are posed often leads pupils to pursue a particular kind of strategy, usually one involving written recording. The directing of attention exclusively to pupils' mental strategies involves a deliberate refraining from indicating that any particular type of response is expected. Thus, pupils are encouraged to feel that what is needed is not the recall of some taught routine but the personal construction of a suitable algorithm from a range of options.

A MONITORING ACTIVITY

Climb the Mountain is an activity in which two players take turns making calculations that will allow them to move a counter up a mountain and down the other side (see fig. 6.1). At each stage the players need to consider the "distance" (difference) between the number their counter is currently on (departure number) and its next resting place (target number) indicated on the playing board. Thus, the context provides an opportunity for the generation of a variety of language forms:

- "How far is it to . . .?"
- "How much farther to . . .?"
- "What's the distance between . . . and . . .?"
- "What's the difference between . . . and . . .?"

In the past, our desire to discover pupils' competence at calculating difference has tended to focus on their performance in carrying out a vertically arranged subtraction algorithm, usually irrespective of the size of the numbers concerned. Thus, the same routine was considered equally appropriate for solving the following:

$$\begin{array}{r} 24 \\ -17 \\ \hline \end{array} \quad \text{and} \quad \begin{array}{r} 3504 \\ -2623 \\ \hline \end{array}$$

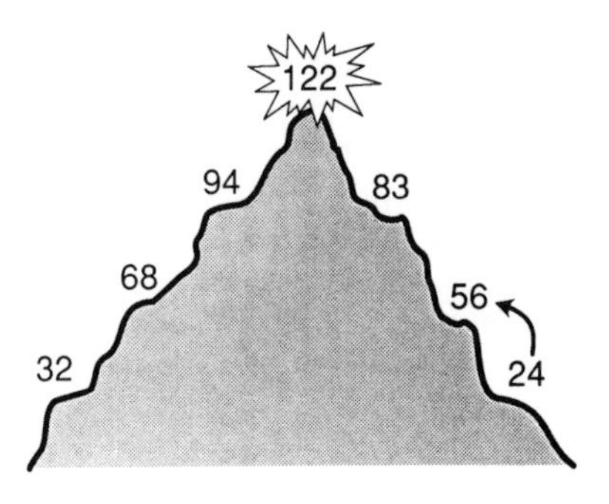

Fig. 6.1

Asking children to describe their algorithms as they play Climb the Mountain strikingly reveals their preference for a mental approach rather than pencil-and-paper methods.

CHILDREN'S CALCULATING STRATEGIES

Digit-by-Digit Comparison

When Gemma (aged eight) tackled the problem of reaching the target number 56 from 24, she explained that she had compared the tens digits in the two numbers, saying to herself, "From 20 to 50 is 30." Then, comparing the unit digits, she said, "From 4 to 6 is add 2." The rules of the game required her to synthesize those two operations into a single operation, the outcome of which would be a calculator algorithm of the following sort:

[+] ? [=]

Eight-year-old Rebecca's explanation for the move from 32 to 68 was very similar, but with the interesting distinction that when the digits were compared (fig. 6.2), she turned to subtraction for the mental operation, saying:

> 8 take away 2 is 4(!), and
> 6 take away 3 is 3, so it's
> add 34.

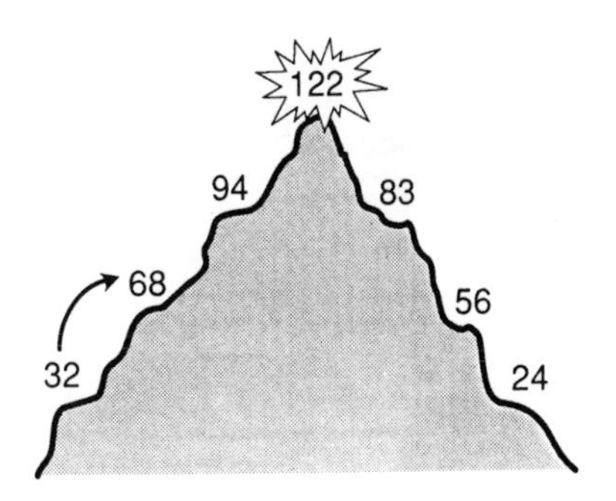

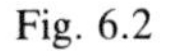

Fig. 6.2

This "digit by digit" comparison approach is a popular one. Many children encounter problems, however, when they attempt to use that strategy regardless of the size of the digits in the pair of numbers. For example, if they attempt to find the move from 68 to 94, applying the digit-by-digit routine results in the following calculator algorithm:

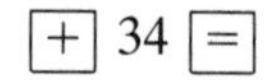

The Climb the Mountain activity can therefore reveal misconceptions about rules for calculating differences, such as "take the smaller number from the bigger." The game is equally helpful in identifying pupils' developing understanding of place value, as shown by their ability to construct strategies that involve counting in tens.

Counting in Tens

John (aged nine) bridged the gap between 56 and 83 (fig. 6.3) by counting on from the departure number in tens:

56 . . . 66, 76, 86

Then he subtracted 3 to arrive at the calculator algorithm:

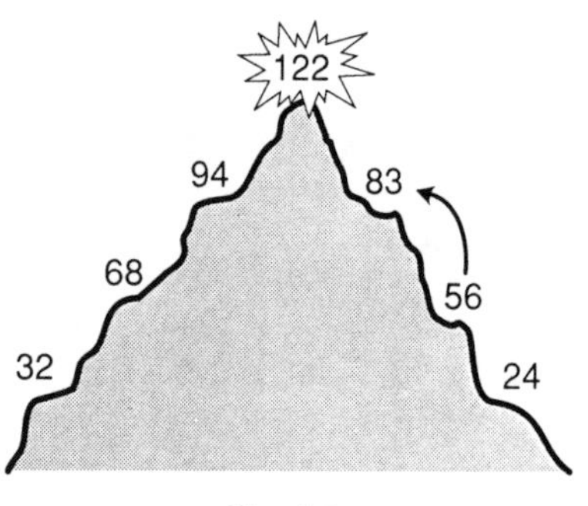

Fig. 6.3

Number Bonds

Another important indicator of pupils' expanding range of skills is their awareness that known number bonds can reduce the burden of counting, whether in ones or in tens. For example, Climb the Mountain enables a child to reveal that she knows that 53 to 93 is "add 40" without counting in tens on her fingers. Eight-year-old Lauren showed her ability to recall and use the fact that 25 + 25 is 50 when she attempted to calculate the difference between 24 and 56. She constructed the calculator algorithm [+] 32 [=], saying to herself that "added to the 25 must be 1, because it's 24, and 6, because it's 56."

APPLYING MATHEMATICAL KNOWLEDGE

The informality of the Climb the Mountain activity allows pupils to respond in idiosyncratic ways that are characteristic of their own mental strategies. Through playing the game, pupils can test their strategies of mental calculation, particularly their developing awareness of place value and their ability to apply their skills of adding or counting in tens. As helpful as this activity can be for assessment purposes, on its own it cannot provide the learning experiences pupils need to develop their understanding of place value. Examples of such activities are suggested below. What Climb the Mountain can do, however, is sharpen the procedures that can transform inert conceptual knowledge or skills into usable and effective calculating strategies. Rote teaching of computational routines has seriously overlooked the need to help children forge links between these two distinct areas.

THE ROLE OF THE CALCULATOR

The preceding examples demonstrate how the calculator can supply children with the rationale for synthesizing a series of stages (or suboperations) into a single operation. For example:

$$37\ (+10) = 47\ (+10) = 57\ (+10) = 67\ (-1) = 66$$

as

$$37\ [\ +\ 29 = \ ]$$

The calculator also gives instant feedback to the pupils, allowing the activity to proceed without a teacher's intervention.

Helping the Development of Invented Strategies

The mental calculation strategies performed by people who handle numbers with confidence show a remarkable variety of nonstandard procedures. Developing a conceptual understanding of number, therefore, necessitates a teaching strategy that lets pupils choose. The calculator can be a very supportive tool for children in their early explorations of numbers, particularly in encouraging a flexible response. Since constructivists generally reject the practice of direct teaching as the major strategy for developing awarenesses, a premium is put on those activities that enable children to extract meaning for themselves, in their own way, and in their own time. The repetitive element offered by games and investigations can effectively furnish that opportunity.

Going Places

Children who are still discovering number notation need experiences that allow them to see, in a gradual and progressive fashion, how numbers grow. The activity Going Places is a good example of such an activity. The teacher offers three pupils an operation die, a calculator, and a recording sheet with the instruction, "Choose a starting number and find out what happens when you get to the end of the sheet."

Each child has a specific role to play in this activity. The die thrower tells the other two which operation has come up. An appropriate range of die operations might be the following:

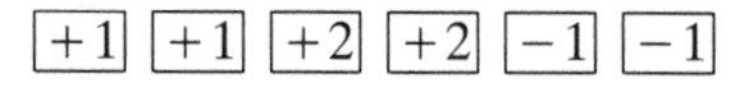

The die thrower and the recorder try to predict the effect of performing the operation on the starting number. The calculator operator presses the appropriate keys and informs them if they are correct. The die operation is then entered onto the recording sheet followed by the new number, ready for the next throw of the die:

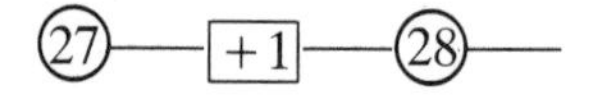

This kind of open-ended context is also well suited for helping children discover for themselves the principal aspects of place value. The teacher can change markings on the die to

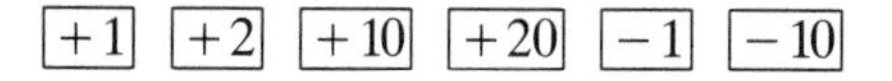

and later to

+2 +10 +20 +100 −1 −10.

Through discussion and by changing roles in the activity, pupils can learn to predict the effect of adding and subtracting hundreds, tens, or units as a single operation.

A minor structural change in the activity, however, can require pupils to replicate the strategy that we saw was characteristic of attempts to solve difference problems in Climb the Mountain. For example, two dice are thrown simultaneously, one with addition functions (e.g., +1, +2, +10, +20, +100, +200) and the other with subtraction functions (e.g., −1, −1, −2, −2, −10, −10). The task of the calculator operator now is to enter—separately—both operations. The task of the recorder is to combine the two operations into a single one (e.g., +20 and −1 as +19) and enter that into the space provided on the recording sheet. All participants in the activity then attempt to predict what the outcome will be when the = key is pressed. Thus the activity may help them to construct the idea that skills they have already learned—adding and subtracting a ten or a unit—can be combined to yield a tool for "bridging the gap" (or finding the difference) between two numbers that are not so conveniently separated. The pupils are being encouraged to carry out precisely the sort of operations on number that will underpin the development of successful mental strategies such as those described earlier.

Steering the Number

The calculator performs a rather different role in the game Steering the

Number. In this activity, a group of children individually (or in pairs) attempt to steer a number, represented in concrete form by rods (tens) and cubes (units), toward one of their own numbers. The preparatory stage of the game consists of each player randomly generating ten different numbers between 1 and 99. Those numbers are written down in a list. A starting number is then produced by throwing two dice. The outcome is interpreted as a two-digit number that is then shown by placing rods and cubes onto a base board (fig. 6.4).

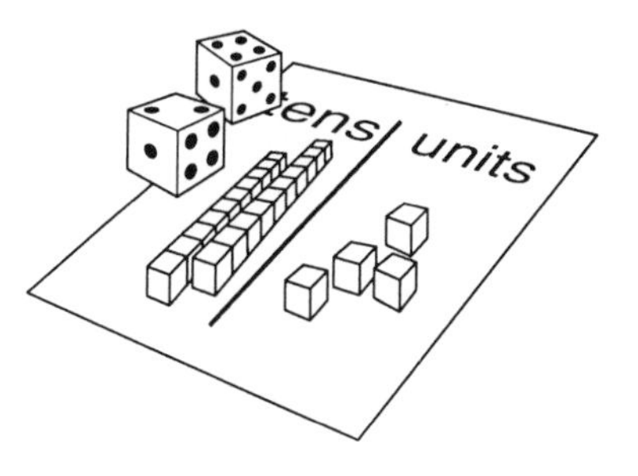

Fig. 6.4

Players now take turns to change the board in a way that will produce a number on their own list, which they then cross off. If in changing the board a player makes a number on someone else's list, that number can also be crossed off. Two rules govern how a number may be changed: (1) Any number of elements may be added to, or subtracted from, the board, but (2) only one side of the board may be altered (*either* the tens *or* the units).

Each action must be accompanied by a "commentary," such as the following: "I'm adding five units." "I'm adding three tens." "We're subtracting six units." "We're subtracting one ten." The calculator can be introduced when the idea behind the activity has been firmly established and the teacher's continual presence is no longer required. Its introduction provides a valuable link between the action taking place, that is, the adding or removing of a number of rods or cubes, the words that name the number of items, and the place value of the digit. For example, accompanying the words "we're adding three tens" is the calculator algorithm [+] 30 [=]. A child who forgets or neglects to interpret the handling of three ten-rods as "30" soon discovers, on pressing [=], that his or her calculator number does not correspond to the number displayed on the board. In this sense the activity is "self-correcting."

NEED FOR CONSISTENCY

Whereas Climb the Mountain is not designed to help children actually build calculating skills but to provide a suitable context for applying them,

Going Places and Steering the Number are intended for skill building. They all form part of a program (Shropshire Mathematics Centre 1990) of awareness and skill-building activities designed to open up multifaceted options for calculating rather than to train students in a single, formal algorithm.

Although it is likely that in the short term teachers will add such activities to their repertoire on an ad hoc basis, the flexible, informal mental calculation activities do sit rather uncomfortably alongside parallel attempts to instruct pupils in specific computational routines. The existence of the pocket calculator thus presents a challenge to teachers to offer pupils a consistent methodological approach—one that not only takes advantage of the technology itself but also takes full account of the powers that all pupils possess.

TEACHERS' RESPONSES

As part of an in-service process, I was able to assess the response to these ideas by a small group of teachers of eight-to-ten-year-olds who agreed to a trial of a program of activities including those described above (Sugarman 1989). Over a six-week period all the teachers were able to refer to specific areas of progress that they had observed with individual pupils. In such a short time I had not anticipated that much individual pupil progress would be noticeable. What I had been particularly keen to discover was whether teachers were able to monitor their pupils' development through the Climb the Mountain activity. In fact, all the teachers were able to report actual development in strategies used by certain pupils, apparently as a result of the program of activities. Among their observations were these:

- "Alison progressed from relying on counting in ones to counting in tens, for instance, in moving from 7 to 37, 7 . . . 17, 27, 37 . . . = 30."
- "Louise was able to climb from 41 to 70 by counting 51, 61, 71, then subtract 1, when at the start, she would only guess with very little idea of how to go about it, or count in ones."
- "Her mental calculations became very fast, for example, an immediate calculation of 27 + 30."

The teachers were also able to recognize that these activities were part of an approach different from that of the commercially published schemes in use in most schools. Referring to the package of activities from the Shropshire Mathematics Centre (1990), one commented:

> It shows children there are a *range* of perfectly valid strategies that can be used to solve problems—none more "right" than others but one of which may be easier to a particular child. It gives more chance of success if there are several ways than if one method is taught.

All teachers responded positively to the way in which the activities were received by their pupils. They recognized, however, that the interactive nature of the approach may impose heavy demands on children's social skills, which in the short-term may need careful attention and perseverance. The use of the calculator was felt to be a strong motivating factor.

CONCLUSION

I have attempted to show that the change toward a more concept-based approach to the school mathematics curriculum, already overdue but made more urgent by the arrival of the inexpensive pocket calculator, can be effected by a constructivist approach to teaching methodology. Such an approach is exemplified by the two concept-building activities described and by the use of appropriate assessment activities, such as Climb the Mountain. Through the use of such informal activities, it is possible to raise to the surface the mental processes of children and to gain insight into their creative mathematical powers. I have argued that an approach like this one can avoid the element of instruction or the imposition of routines during the primary phase of schooling. In so doing, children are "functioning as mathematicians" in a way that cannot be said to be true of traditional approaches.

REFERENCES

Ashfield, David. "Counting in Ones." *Mathematics in School* 18 (January 1989): p. 25.

Department of Education & Science. *Mathematics in the National Curriculum.* London: Her Majesty's Stationery Office, 1989.

Ginsburg, Herbert. *Children's Arithmetic: How They Learn It and How You Teach It.* Austin, Tex.: PRO-ED, 1982.

National Curriculum Council. *Non-Statutory Guidance.* London: The Council, 1989.

Sugarman, Ian. "A Constructivist Approach to Developing Children's Mental Strategies for Calculating Difference." Master's thesis, University of Keele, 1989.

Shropshire Mathematics Centre. *Developing a Feeling for Number.* Shrewsbury: The Centre, 1990.

7

Decimals and Calculators Make Sense!

DeAnn M. Huinker

IN THE whole number strand of school mathematics, much time is spent developing meaning for the numbers and place value prior to computation. But with decimals, instruction often focuses almost immediately on computation. Students are given little, if any, time to develop meaning for decimal symbols. When students do not know what the symbols mean, they have no way of figuring out why the rules work and usually have no choice but to memorize rules and rely on rote manipulations of symbols when computing with decimals (Hiebert 1985, 1987). This leaves students unable to judge whether their answers are reasonable (Wearne and Hiebert 1988). For students to understand what they are doing, they need to develop number sense and operation sense of decimals.

This article explores ways to teach those decimal concepts using calculator activities to help students develop understanding. The activities were written using a TI-108 calculator, but they work well on many other calculators that have automatic constant functions for all four operations. With slight modifications the activities can be used with the TI Math Explorer and even with scientific calculators that have a special constant key.

NUMBER SENSE

A student with strong number sense possesses well-developed knowledge of—

- meanings for numbers and their representations in symbols;
- relationships among numbers;
- relative magnitudes of numbers;
- relative effects of operating on numbers;
- referents for numbers in common objects and situations (National Council of Teachers of Mathematics 1989).

Decimal number sense first involves having meaning for decimal symbols. To develop well-understood meanings, students need many concrete experiences connecting quantities or models of tenths, hundredths, and beyond to their oral names and to their symbols. Activity 1 describes an initial activity for helping students make connections among the models, oral names, and symbols for tenths. A similar activity for hundredths could be accomplished by using a ten-by-ten square or base-ten blocks. This activity uses the calculator constant function for addition. It is important for students to be familiar with the constant function before doing this activity.

Activity 1: Connecting Quantities, Oral Names, and Symbols

Each pair of students will need some paper strips marked in tenths, a calculator, and crayons. Pretend that each paper strip is a granola bar. One student colors in one-tenth of the bar to indicate that part was eaten and states the amount shaded while the other student presses [+] .1 [=] on the calculator to see the symbolic representation on the display and to prepare the calculator to count by tenths. Continue to color in the tenths, state the amount shaded, and press the [=] key to connect the symbols to the quantities. Stop at five-tenths and ask, "Now, you've eaten five-tenths of your bar. Is there another name that we can use to tell how much we ate?" Have the students fold their paper strips in half to see that five-tenths is the same amount as one-half. Stop at ten-tenths and discuss how this is another name for one bar. Keep the activity going by using more paper strips while stopping at 1.5, 2, 2.5, and so on, to discuss the relationship of the number of tenths eaten to one-half and to the whole bars.

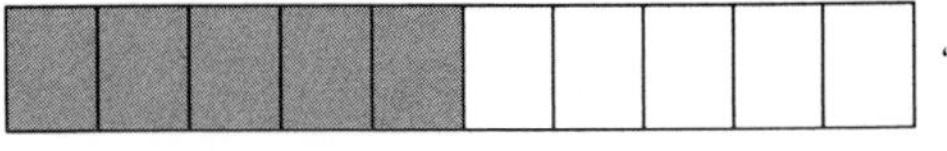

"Five-tenths is the same as one-half."

The place-value system for recording decimal numbers is an extension of that for whole numbers. The main ideas are the same; the value of a digit depends on the position in which it is placed, and the value of any position is ten times larger than the value of the position to its immediate right. When the whole-number numeration system is extended to also express fractional quantities as decimals, calculator activities can help students to see the ten-times-larger pattern and to observe that the reverse is also true: that the value of any position is ten times smaller than the value of the position to its immediate left. The pattern can be extended to the right of the ones place giving tenths, hundredths, thousandths, and so on. Use Activity 2 to reinforce this understanding.

Activity 2: The Place-Value Mover

As the students enter this code into their calculator, [C] 10 [×] 3 [=] [=] [=] [=] [=] [=] [÷] 10 [=] [=] [=] [=] [=] [=], have them predict what the calculator will display before each [=] key is pressed. Then have the students read the number. Continue predicting and reading the number as it grows from 3 to 3 000 000 and as it shrinks back to 3. Now ask the students to predict what will happen if the [=] is pressed again. Do it and then discuss why it displayed 0.3. Continue to predict and press the [=] key until you reach "3 millionths." Now discuss the symmetry around the ones place (*not* the decimal point!) in our numeration system.

Use this calculator code with the TI Math Explorer: [C] 3 [×] 10 [=] [=] [=] [=] [=] [=] [÷] 10 [=] [=] [=] [=] [=] [=].

It is important to help students understand the symmetry about the "ones" place—tens on one side, tenths on the other; hundreds on one side, hundredths on the other; and so on. A common error is the notion that there is symmetry about the decimal point rather than the ones place. Students may think that 7 is in the "oneths" place in 0.7. The decimal point merely determines the ones place, separating the whole units from the fractional part. Students enjoy playing Wipe Out (see Activity 3) as a challenge to their knowledge of place value. Discuss and compare the symmetric place values as you play this game. For example, if a student wanted to wipe out the 3 in 1359.247, discuss how to wipe out 3 hundreds (subtract 300) as compared to 3 hundredths (subtract .03).

Activity 3: Wipe Out

Have the students enter a number on a calculator such as 7235.498 and then challenge them to wipe out the 4 in a single operation without changing any of the other digits. To accomplish this, the students must realize that the 4 is in the tenths place and that therefore they must subtract 0.4 to wipe out the 4. To wipe out the 2, the students must realize the 2 is in the hundreds place, so they must subtract 200 to wipe out the 2. Have the students share their strategies. Once the students have learned the game, they can easily play it in small groups.

Decimal number sense also involves developing relationships among decimals, such as equivalent decimals and relationships to other decimals. A knowledge of equivalent decimals includes understanding that "3 tenths" is the same quantity as "30 hundredths" and that 0.3 is equal to 0.30. This knowledge helps students make more sense of the symbols and be more flexible with computational procedures (Payne, Towsley, and Huinker 1990). Here are some more examples of useful relationships among equivalent decimals: 10 tenths is the same as one whole unit; 40 hundredths is the same as 4 tenths; 3 tenths 7 hundredths is the same as 37 hundredths; and 15 tenths is the same as 1 whole unit and 5 tenths. Activity 4 emphasizes relationships among equivalent decimals as students count by hundredths.

Activity 4: Counting by Hundredths

Have the students count aloud by hundredths as they also make the calculator count by hundredths: [C] [+] .01 [=] [=] [=] [=] [=] . . . . After 0.09 is displayed, ask, "What do you think the calculator will show us next? (10 hundredths) What did it show us? (0.1) Is the calculator wrong?" Use base-ten blocks to show that 10 hundredths is the same amount as 1 tenth. Continue counting, saying both "20 hundredths" and "2 tenths" when 0.2 is displayed, and so on, and discuss why these are the same amount.

The ability to recognize the relative magnitude of decimals contributes most to making sense of these numbers and using them with understanding. Students need to have a sense of those decimals that are close to zero, close to one-half, and close to one. For example, 0.00045 is a very small quantity that is close to zero, 6.87 is close to seven whole units, and 0.49023 is close to one-half. Older students should also have a sense of those numbers that are close to one-fourth and three-fourths. Students can also be challenged to recognize decimals that are close to but larger than (and close to but smaller than) each of these familiar quantities. Discussions of the relative magnitude of numbers should be incorporated into all work with decimals, from initial concept development to computation.

OPERATION SENSE

Students need experiences developing quantitative understanding of decimals before they are asked to compute with them. When students can make sense of decimals as numbers, they are ready to develop decimal operation sense. Think about the following questions to determine whether you have operation sense of decimals. Can you describe real-world situations in which you would add, subtract, multiply, or divide with decimals that do and do not involve money? Will the result of 0.43 times 86.34 be larger than 86 or smaller than 86? Can you state the previous problem (and others) in simpler terms to estimate the amount? Check your thinking with your calculator. Will the answer to 0.927 divided by 3.0146 be about 30, about 3, about 0.3, or about 0.03? How can you tell? Check your thinking with your calculator.

A student with well-developed operation sense has a knowledge of—

- mathematical operations that model operations in real-world situations;
- the fundamental properties of operation;
- relationships among operations;
- the effects of each operation on a pair of numbers (National Council of Teachers of Mathematics 1989).

Addition and Subtraction

Ask students to tell stories, real or imaginary, about situations in which they or someone else needs to add or subtract with decimals. These stories can be simple word problems or elaborate stories with a setting and a plot (see Bush and Fiala [1986]). Let the students use calculators as they plan and solve one another's problems, since calculator use allows students to concentrate on the use of decimals and the meanings of the operations in different contexts. Problem posing is also an effective method of evaluating students' number and operation sense of decimals.

Rather than memorizing what to do with the decimal points, students should use their understanding of the quantities and knowledge of the effect of operating on a pair of numbers to reason through the placement of the decimal point. To reason through the first problem in Activity 5, students should realize this is about 7 plus 96, so the answer is about 100. Now they can place the decimal point with understanding. This activity could also be connected to real-world problems or problems posed by the students.

Activity 5: WANTED—Decimal Points

Inform the students that the following answers have lost their decimal point! Have the students estimate to place the decimal point in each answer to make it correct. Then check with a calculator.

7.2588642 + 96.365224	=	1 0 3 6 2 . . .
0.203685 + 853.08952	=	8 5 3 2 9 3 . . .
79.65632 − 52.92415	=	2 6 7 3 2 . . .
6.756821 − 6.4365337	=	3 2 0 2 8 . . .

Results from national assessments (Carpenter et al. 1981a, 1981b; Kouba, Carpenter, and Swafford 1989) have documented the difficulties students have in solving decimal computational exercises that are presented in a horizontal format and involve decimals expressed in unlike units. Activity 6 helps students make sense of these trouble spots.

Activity 6: Trouble Spots

Have students work in small groups of four. One student uses base-ten blocks, another uses paper and pencil, the third uses a calculator, and the fourth is the reader. The reader verbally presents the trouble-spot problems to be solved. Have the group members compare what is done and clarify any discrepancies in results. The students switch roles after every other problem. Here are some examples of trouble spots.

"Five minus two-tenths equals"
"Six minus two-hundredths equals"
"Ten minus five-tenths equals"
"Twelve minus five-hundredths equals"
"Three-tenths minus seven-hundredths equals"
"Seventy-two hundredths minus four-tenths equals"

Multiplication and Division

Multiplication problems that require decimals usually involve continuous quantities. For example: Glen bought 3 packages of sausage. Each package weighed 1.4 pounds. How much meat did Glen buy? ($3 \times 1.4 = 4.2$. Glen bought 4.2 pounds of meat, or 4 whole pounds and 2 tenths of another pound.) The next example involves finding the product with two continuous quantities: The room is 6.2 meters wide and 8.7 meters long. What is the area of the room? ($6.2 \times 8.7 = 53.94$, so the area is almost 54 square meters.)

A real-world situation for division might involve sharing some quantity among a specific number of people to find out how much each person will get (sometimes called "partitioning"). For example, if eight people are going to share a 2-liter bottle of soda pop, how much will each person get? ($2 \div 8 = 0.25$. This means that each person will get 0.25, or one-fourth, liter of soda pop.)

Another type of real-world situation involves the repeated subtraction of equal parts (sometimes called "measurement"). For example, how many smaller pieces of wood that are each 0.4 meter long can be cut from a piece of wood that is 6 meters long? ($6 \div 0.4 = 15$. This means that 15 smaller pieces can be cut.) The posing and illustrating of real and imaginary problems contributes to operation sense and is exciting and fun for students when they can use calculators to solve their own and one another's problems.

The rules for memorizing where to place the decimal point in multiplication and division exercises are often forgotten or confused because the students have no understanding of why these rules work. Students should be encouraged to use their understanding of the quantities and of the operations to reason through the placement of the decimal point. This requires students to make estimates about the effect of operating on a pair of numbers. Unfortunately, many students' estimation ability is not well developed. On a national assessment, only 21 percent of the thirteen-year-olds were able to make a good estimate (or a lucky guess) of 3.04×5.3 (Lindquist et al. 1983). Activity 7 encourages the use of estimation to place decimal points in products and quotients. Students should be encouraged to think about the first problem as "This is about 4 times 8, so the answer is about 32." The last division problem is challenging to think about. It requires 13.654229 to be thought of as "about 136 tenths" and then this is to be shared among 90 people, so each person will get at least one tenth.

Activity 7: WANTED—Decimal Points

Inform the students that the following answers have lost their decimal point! Have the students estimate to place the decimal point in each answer to make it correct. Then check with a calculator.

3.9854×8.036582	=	3 2 0 2 8 9 9 . . .
674.3652×0.975036	=	6 5 7 5 3 . . .
$49.873215 \times 0.531422$	=	2 6 5 0 3 7 . . .
$39.87636 \div 2.138577$	=	1 8 6 4 6 . . .
$13.654229 \div 89.365246$	=	1 5 2 7 9 . . .

Activities 8 and 9 encourage the development of operation sense through problem solving as students make estimates and study the effects of operating with decimals. Students investigate the effect of multiplying with decimals in "Hit the Target" and study the effect of dividing with decimals in "Discover the Secret Number."

Activity 9 develops division sense and a sense of relative magnitude. It is important for students to understand clearly how the constant function for division works before doing this activity. As a test, enter this code into your calculator and note what happens: 10 [÷] 2 [=] 28 [=] 35 [=] 42 [=] 15 [=]. After successive [=] keys, the display will show 5, then 14, then 17.5, then 21, and then 7.5. The calculator is repeatedly dividing the number in the display by 2, or is using 2 as the division constant. Try it with a different division constant.

Activity 8: Hit the Target

In pairs, one student selects a target range, for example, 870 to 890, and the other chooses a start number, for example, 53. The first student then enters the start number, the [×] key, an estimate, and the [=] key. If the displayed number falls within the target range, the student wins. If not, the calculator is passed to the other student, who (without clearing the calculator!) enters the [×] key, an estimate, and the [=] key. Keep playing until the display falls within the range. Here is an example of a short game. It usually takes much longer.

Player 1 enters the start number: 53

		Estimate		*Display*	
Player 1	[×]	22	[=]	1166	
Player 2	[×]	0.9	[=]	1049.4	
Player 1	[×]	0.7	[=]	734.58	
Player 2	[×]	1.2	[=]	881.496	The winner!

In Activity 9, the constant function for division is being used. In the example given, the secret number is 32 and the first guess entered was 76. The student should think, "76 divided by the secret number is 2.375. This

Activity 9: Discover the Secret Number

In pairs, one student selects a secret number between 0 and 100, for example, 32, and then hides it in the calculator by pressing 32 [÷] 32 [=]. The display now shows the digit 1. Pass the calculator to the other student, who tries to discover the secret number by using only the number keys and the [=] key. This student should keep track of the guesses and the resulting display, noting whether the display is greater than 1 or less than 1. When the display is exactly 1, the secret number is discovered. Here is an example of a search for the secret number.

Guess		*Display*	*Note*
76	[=]	2.375	Greater than 1 because . . .
40	[=]	1.25	Greater than 1 because . . .
30	[=]	0.9375	Less than 1 because . . . (Also, getting closer because . . .)
34	[=]	1.0625	Greater than 1 because . . .
32	[=]	1.	Exactly 1. I discovered the secret number! It's 32.

means that 76 is more than twice the size of the secret number, or the secret number is about one-half of 76." For the guess of 30, the student should think, "30 divided by the secret number is 0.9375. This means that the secret number is larger than 30 because the answer is less than 1."

CLOSING COMMENT

Mathematics, even decimals, can make sense, and calculators are a valuable tool in helping students make sense of the numbers they use. As students develop decimal number sense and operation sense, they also develop a sense of themselves as confident and meaningful users of decimals.

REFERENCES

Bush, William S., and Ann Fiala. "Problem Stories: A New Twist on Problem Posing." *Arithmetic Teacher* 34 (December 1986): 6–9.

Carpenter, Thomas P., Mary Kay Corbitt, Henry Kepner, Mary Montgomery Lindquist, and Robert E. Reys. "Decimals: Results and Implications from National Assessment." *Arithmetic Teacher* 28 (April 1981a): 34–37.

———. *Results from the Second Mathematics Assessment of the National Assessment of Educational Progress.* Reston, Va.: National Council of Teachers of Mathematics, 1981b.

Hiebert, James. "Children's Knowledge of Common and Decimal Fractions." *Education and Urban Society* 17 (1985): 427–37.

———. "Research Report: Decimal Fractions." *Arithmetic Teacher* 34 (March 1987): 22–23.

Kouba, Vicky L., Thomas P. Carpenter, and Jane O. Swafford. "Number and Operations." In *Results from the Fourth Mathematics Assessment,* edited by Mary Montgomery Lindquist, pp. 64–93. Reston, Va.: National Council of Teachers of Mathematics, 1989.

Lindquist, Mary Montgomery, Thomas P. Carpenter, Edward Silver, and Westina Matthews. "The Third National Mathematics Assessment: Results and Implications for Elementary and Middle Schools." *Arithmetic Teacher* 31 (December 1983): 14–19.

National Council of Teachers of Mathematics. *Curriculum and Evaluation Standards for School Mathematics.* Reston, Va.: The Council, 1989.

Payne, Joseph N., Ann Towsley, and DeAnn Huinker. "Developing Fraction Knowledge." In *Mathematics for Young Children,* edited by Joseph N. Payne, pp. 175–200. Reston, Va.: National Council of Teachers of Mathematics, 1990.

Wearne, Diane, and James Hiebert. "Constructing and Using Meaning for Mathematical Symbols: The Case of Decimal Fractions." In *Number Concepts and Operations in the Middle Grades,* edited by James Hiebert and Merlyn Behr, pp. 220–35. Reston, Va.: National Council of Teachers of Mathematics, 1988.

8

Concept Development and Problem Solving Using Graphing Calculators in the Middle School

Charles B. Vonder Embse

THE ideal calculator environment for middle school students to learn prealgebra concepts, to explore patterns and processes, and to solve problems should be one in which multiple inputs and outputs of the calculator are shown clearly. It should be an interactive environment that allows students to explore and experiment in ways that help them better understand various parts of a process, pattern, or problem situation. It should be an affordable environment. It should be a portable environment allowing the calculator to be used in class, at home, or anywhere the student chooses. It should be an environment that fosters the ideas of mathematical reasoning, connections, and communication, as outlined in the *Curriculum and Evaluation Standards for School Mathematics* (National Council of Teachers of Mathematics 1989), through the integration of numerical and graphical representation. In many respects this ideal calculator environment sounds more like a computer environment. In fact, the intersection of calculator and computer technology, the graphing calculator, is this ideal environment for teaching and learning mathematics in the middle grades.

The large screen display, graphics capability, and, most important, the exploratory functions of graphing and multiline display calculators afford the middle school student and teacher opportunities to investigate, compare, and explore concepts and problem situations in better ways than when using standard hand-held calculators or no technology at all. The NCTM *Standards* expresses the new vision of mathematics instruction, K–12, which stresses problem solving, communication, reasoning, and mathematical connections throughout the curriculum. Multiline or graphing calculator technology can be the tool that helps students and teachers realize this new vision of school mathematics.

THE LARGE SCREEN DISPLAY: AN UNDERSTANDABLE WORKING ENVIRONMENT

Many students' first encounter with technology in the mathematics classroom is the use of a standard hand-held calculator. At the middle school level, students learn to use a scientific calculator, which uses algebraic hierarchy. A stumbling block for many novice users is the need to visualize, or hold in memory, a complicated string of numbers, operational symbols, and grouping symbols as they are entered into a calculator. For example, perform the following calculation:

$$3 + 4\,(5 - \frac{6}{7})$$

Figure 8.1 shows the sequence of views that the student sees as this arithmetic computation is entered into a standard scientific calculator.

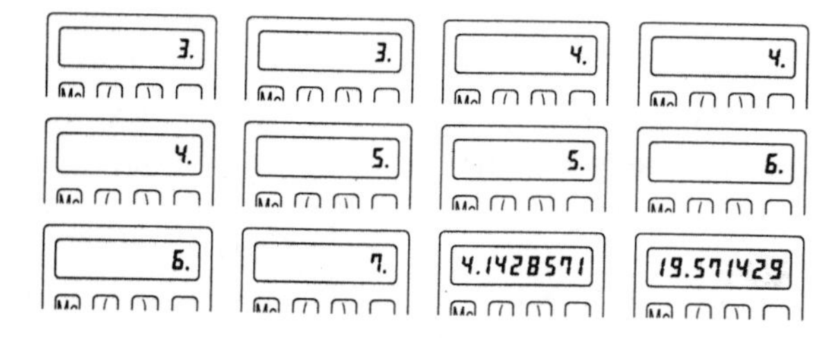

Fig. 8.1. Sequence of views on a standard hand-held calculator

Only the numerical entries and calculated values are shown on the display; the operational and grouping symbols are not shown. Sometimes intermediate answers are shown. In the example above, if the final [=] key is not pressed, the student might think that the answer is 4.1425871 instead of the correct value of 19.571429.

Figure 8.2 shows the same computation entered on a graphing calculator. The large screen functions like a computer screen, showing the entire key sequence and the answer at the same time. The screen display of the keying sequence closely resembles the actual mathematical phrase. Figure 8.3 shows the same example entered without using the multiplication symbol

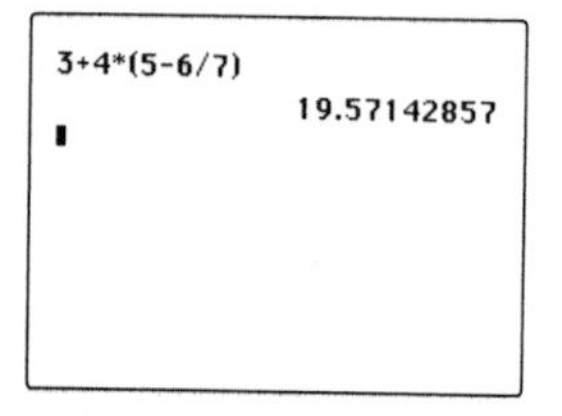

Fig. 8.2. Calculation entered on the large screen of a graphing calculator

Fig. 8.3. Two calculations entered on the screen at the same time

between the 4 and the quantity in parentheses; the calculator understands that juxtaposition implies multiplication. Multiple problems and answers can be shown on the screen at the same time for careful comparison of results. The answer is on the screen at the same time as the keying sequence, thus reinforcing the functional idea of input and output.

Entry errors are another stumbling block when using a standard calculator. Incorrect numerical entries may slip by unnoticed. Incorrect operational and grouping symbols are even more difficult to detect, since these keystrokes are invisible to the student and the teacher. A common question from students is, "I didn't get the same thing you did. What did I do wrong?" When the teacher looks at the final numerical result on the display screen, it is often just a guessing game to determine which incorrect keystrokes the student made. If multiple incorrect keystrokes were made, the puzzle may be impossible to solve. The common response from the teacher is for the student to clear the machine and begin again. Error analysis is often more instructive to the teacher than correct answers. It can help the teacher detect and correct students' misconceptions about the mathematics and help students understand their own error patterns. Without a visual record of the keystrokes used to arrive at an answer, there is limited possibility for error analysis by the teacher or the student.

A multiline display allows a comparison of up to four calculations at the same time. Figure 8.4 shows the screen of a graphing calculator with three versions of the problem above entered at the same time. Removing the parentheses in the last entry results in a different calculation and a different answer.

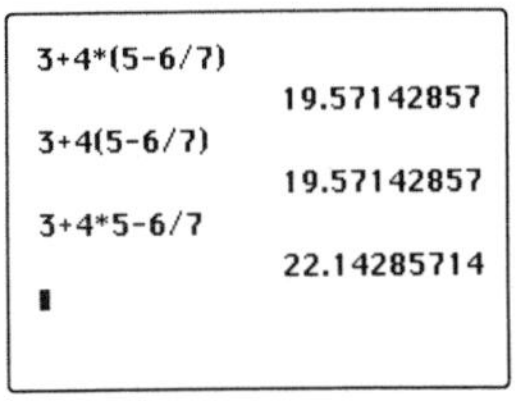

```
3+4*(5-6/7)
                 19.57142857
3+4(5-6/7)
                 19.57142857
3+4*5-6/7
                 22.14285714
▮
```

Fig. 8.4. Multiple problems on the screen at the same time

Entry errors can be seen and corrected immediately by using **Insert** and **Delete** functions to correct or change the input rather than retyping a problem. These editing functions work the same on a graphing calculator as on a computer. When doing numerical problem solving using the "guess and check" problem-solving strategy, one may reenter a single problem ten, twelve, or as many as fifteen times. Each reentry of a problem is an opportunity to make undetected keying errors. Students often get careless when repeating the same operations over and over again; errors can become more frequent. The constant visual record of keystrokes on the screen of a graphing calculator helps students check their entry for accuracy. The editing features also mean that if an error is made, the student can correct the mistaken entry and recalculate.

ALGEBRAIC ORDER OF OPERATIONS

One of the first topics that must be covered when dealing with calculator or computer technology in the mathematics classroom is *algebraic* order of operations. At the middle school level it is appropriate for students to use calculators that have algebraic hierarchy built into their operating systems. The large screen of the graphing calculator allows students and teachers to experiment with various mathematical phrases and develop the order of operations rules. For example, the rule that multiplication is done before addition can be shown by entering the mathematical phrase $2 + 3 \times 4$ on the graphing calculator (see fig. 8.5). Algebraic order of operations performs the multiplication before the addition, resulting in an answer of 14.

If a *sequential* order of operations is used, the result of this computation would be 20. In order to make this phrase result in 20 by algebraic order of operations, the calculator must add 2 and 3 before it multiplies by 4. This is done by using parentheses to regroup (see fig. 8.6).

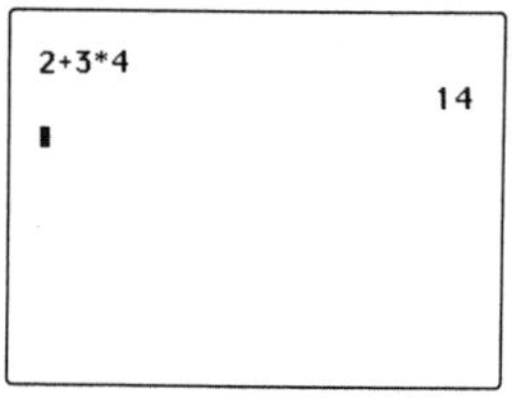

```
2+3*4
            14
```

Fig. 8.5. The calculation "2 + 3 × 4"

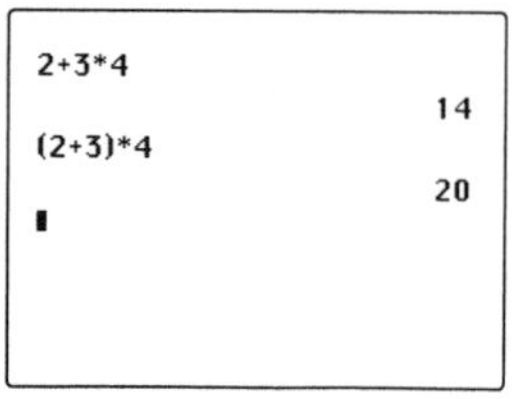

```
2+3*4
            14
(2+3)*4
            20
```

Fig. 8.6. Parentheses to group "2 + 3"

What would happen if we grouped the multiplication factors in parentheses? Figure 8.7 shows the modified phrase entered on the screen.

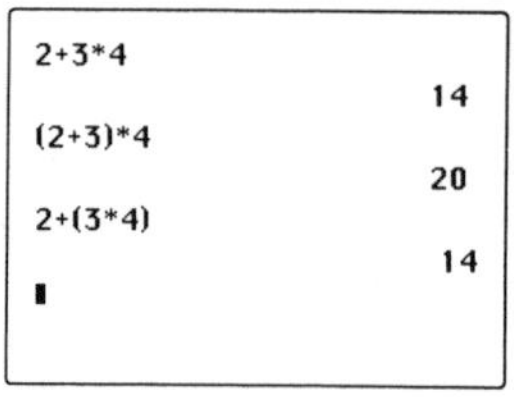

```
2+3*4
            14
(2+3)*4
            20
2+(3*4)
            14
```

Fig. 8.7. Inserting parentheses around the multiplicative factors is not necessary

The parentheses cause the multiplication to be done first, but that symbolic grouping is not needed, since the algebraic order of operations will do the multiplication before the addition. The multiple keying sequences and answers shown on the screen at the same time help students see the connection between the examples, thus enhancing the teaching power of the examples.

The fraction bar is interpreted as a grouping symbol when used in a mathematical phrase. For example,

$$\frac{7 - 3.5}{9 - 2.7}$$

would be entered as seen in figure 8.8. The keying sequence reinforces the idea that the fraction bar is a grouping symbol. When the same calculation is entered again but without the parentheses (see fig. 8.9), the new answer indicates that a different calculation has been done. If students analyze the keying sequence shown on the screen by algebraic order of operations, this second keying sequence really represents the mathematical phrase

$$7 - \frac{3.5}{9} - 2.7.$$

```
(7-3.5)/(9-2.7)
                .5555555556
```

Fig. 8.8. The fraction bar indicates grouping

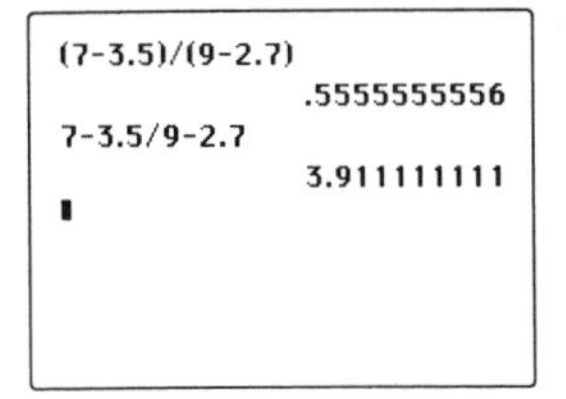

Fig. 8.9. Parentheses omitted

More complicated problems make the graphing calculator's large screen display of the entire keying sequence even more valuable.

PROBLEM SOLVING AND EXPLORATION THROUGH TABLE BUILDING

The large text screen of the graphing calculator allows students to solve problems numerically by repeatedly entering the required calculations with different input values. For example, consider the following problem:

> Many triangles have a height of 7 units. Make a table that shows how the area of such a triangle depends on the length of the base.

Figure 8.10 shows the screen display for calculating the area of a triangle of height 7 and base 1; the area is 3.5 square units. Notice how the screen display shows the numerical entries, the operational symbols, and the computed result all at the same time.

The next line in the table could show the area of the triangle with a height of 7 units and a base of 2 units. If we continue this process using values of 5 and 10 for the size of the base, the screen will look like figure 8.11. The screen displays a table showing the input values of 1, 2, 5, and 10, representing the different values for the bases of the triangles, and the output values for the area of the triangle of height 7 and those bases. If we continue this process, the top entry will scroll off the screen, allowing up to four results to be visible at one time. A clear distinction is made between the input and the calculated output.

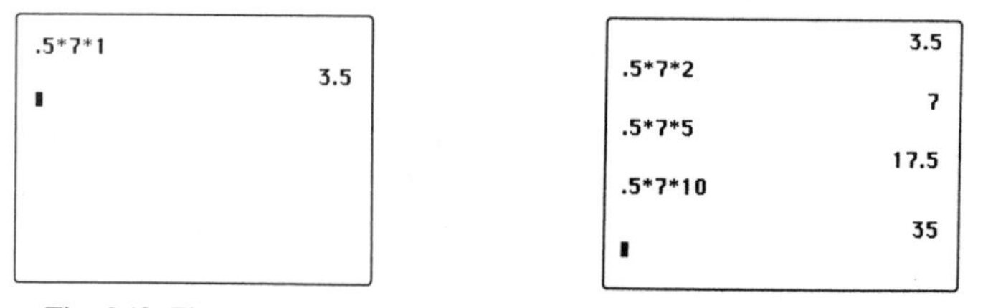

Fig. 8.10. First problem

Fig. 8.11. Display of four rows of the table

Let us now extend this problem:

> For a triangle with a height of 7 units and an area of 95 square units, what is the length of the base?

Without clearing the screen, enter a guess for the length of the base. We can see that a base of 5 units gives an area of 17.5 square units and that a base of 10 units gives an area of 35 square units. By doubling the base, we have doubled the area. Guess 20 units for the base and enter the keying sequence without clearing the screen (see fig. 8.12).

A guess of 20 units is not large enough; try 30 (see fig. 8.13). By observing the values, we see that the base length we are seeking lies between 20 and 30 units, and probably closer to 30, since 105 is closer to 95. A guess of 27 units gives the result seen in figure 8.14. Now we are getting very close. More refinement of the guesses will eventually lead to values that narrow in on the needed base length (see fig. 8.15).

The process of "educated" guessing can continue up to the limits of

```
                7
.5*7*5
             17.5
.5*7*10
               35
.5*7*20
               70
▮
```

Fig. 8.12. Guess 20 for the base

```
             17.5
.5*7*10
               35
.5*7*20
               70
.5*7*30
              105
▮
```

Fig. 8.13. Guess 30 for the base

```
                35
.5*7*20
                70
.5*7*30
               105
.5*7*27
              94.5
▮
```

Fig. 8.14. Try a guess of 27

```
            95.0005
.5*7*27.142
             94.997
.5*7*27.1425
           94.99875
.5*7*27.1427
           94.99945
▮
```

Fig. 8.15. A guess of 27.1427

machine accuracy. At each step the student has three previous steps to guide the next guess. The student is in control of the step-by-step process. Each step emphasizes the dependency of the area on the choice of the base. The height never changes, only the base, as the problem originally stated. Obviously, it would be easy to solve this linear equation for the value of the base by doing **95 ÷ (0.5 × 7) = 27.142857,** but this is not the point of the problem. Rather, the student should make the connection that the value of the base of the triangle is free to vary through many values and that each line in a table is the answer to a question. The line in the table we are seeking is the value of the base that results in an area of 95 square units.

The large screen display allows the student (and the teacher) to see several complete mathematical phrases and the computed answer at the same time. If mistakes occur, they can be edited without reentering the entire problem. Students get a better sense of the relationship between the chosen values and the computed results. Building a table on the screen provides the reference points students need to move through the problem toward a solution without becoming bogged down; the next guess can take place immediately. Once students become proficient at building tables on the screen, this guess-and-check process is even more efficient if the **Replay** function of the calculator is used to enter new guesses.

USING THE REPLAY FUNCTION FOR PROBLEM SOLVING AND EXPLORATION

The **Replay** function of a graphing calculator allows the user (1) to repeat the last complete command line that was calculated, (2) to edit the command, and (3) to recalculate a new result. This feature allows the user to edit the particular part of the command being explored. The following problem of exponential growth shows the use of this function in a problem-solving situation:

> Buzzards Roost Records sells compact disks (CDs) at the regular price of $15 each. The annual inflation rate this year is 8%. If this inflation rate stays constant, how much will a CD cost after 2, 5, 7, and 9 years?

The exponential growth model for this problem is $T = P(1 + R)^N$, where T = the new cost, P = the original cost, R = the annual inflation rate, and N = the number of years of growth. Figure 8.16 shows the screen display of the mathematical phrase representing this problem situation for two years of inflation; CDs will cost $17.50 after two years of inflation at 8 percent a year. Notice that the display on the screen is almost identical to the actual mathematical phrase $15(1 + .08)^2$.

```
15(1+.08)^2
                  17.496
▮
```

Fig. 8.16. First keying sequence

To generate the next line in the table, the user could key in the same problem with the next choice for the number of years of growth. This process could be continued until the table from the problem is complete. However, for complicated keying sequences like this one, this process requires time for all the keystrokes. This slows down the exploration process and detracts from the focus of changing the value in the expression representing the number of years of growth. The graphing calculator's **Replay** function will repeat the last command line calculated. This function rewrites the last command line on the screen and positions the cursor at the end of the expression so that the command can be edited. Once the changes are made using **Overwrite, Insert,** or **Delete** operations, the expression can be evaluated again with the new input values. Figure 8.17 shows the screen after the **Replay** function has been used. The previous command line has been repeated and the cursor is positioned at the end of the line ready to be moved into the expression for editing. In this example, move the cursor onto the exponent 2 and overstrike a 5. Press ENTER to see the new result (see fig. 8.18). After five years, a CD will cost about $22.04.

```
15(1+.08)^2
                  17.496
15(1+.08)^2▮
```

Fig. 8.17. **Replay** function used

```
15(1+.08)^2
                  17.496
15(1+.08)^5
             22.03992115
▮
```

Fig. 8.18. A new value calculated

The **Replay** function may be used as often as needed to solve a problem. The following problem extension can be solved by the "guess and check" method using the **Replay** function:

> If the inflation rate stays at 8% a year, how long will it take until the price of CDs is greater than $100?

Notice that after five years of inflation at an annual rate of 8 percent, the cost of a CD has increased about $7. Let's try a guess of 10 years (see fig. 8.19).

After ten years the cost has more than doubled to approximately $32.38. Let's try 15 or 20 years. Figure 8.20 shows the screen of the graphing calculator after several guesses; after twenty-five years, the cost of a CD will exceed $100 if inflation remains at 8 percent a year.

```
15(1+.08)^2
                17.496
15(1+.08)^5
           22.03992115
15(1+.08)^10
           32.38387496
▮
```

Fig. 8.19. After 10 years of inflation

```
           32.38387496
15(1+.08)^20
           69.91435716
15(1+.08)^25
           102.7271279
15(1+.08)^24
           95.11771106
▮
```

Fig. 8.20. After 25 years

The most important aspect for students to notice about this problem is that the total cost is a *function* of the number of years of growth. The amount of time that we are investigating is free to change. When the function is written as an equation, $T = 15\ (1 + .08)^N$, the variable N must represent many values, like the lines in our table. By allowing students to repeatedly change the value for N in the expression, we focus emphasis on what the problem is really asking, "How many years until CDs cost more than $100?"

Consider a further extension:

> If the inflation rate had been 6% a year, what would the cost of CDs be after 25 years?

What is your guess? Use the **Replay** function to edit the percent of inflation. Will the CDs cost less than $100 after twenty-five years? How much less? How long will it take the price of CDs to exceed $100 at an inflation rate of 6 percent a year? Can you make a general rule?

The algebraic solution to this problem involves logarithms, but presenting and exploring the concept of exponential growth should not depend on students' knowing about logarithms or how to manipulate complex algebraic expressions. The guess-and-check problem-solving strategy presented in this example is accessible to middle school students. Guess and check is a generic problem-solving strategy that can be used in many situations, and given modern technology like the graphing calculator, this process is also very efficient. Guess and check causes students to refer continually to the original problem situation and helps develop intuition about mathematical processes and number sense. It allows students to solve real problems of interest to them.

DEVELOPING THE CONCEPT OF A VARIABLE

Many students experience difficulties in algebra and other higher mathematics courses because they never really understand the concept of a variable and how variables allow the generalization of numerical processes. Often students are introduced to the concept of variable intertwined within the concept of an equation in a definition like the following:

> A variable is a letter that stands for a number.
>
> In the equation $4x + 5 = 13$, x is a variable.

This kind of definition obscures the real meaning of a variable. A variable is indeed a symbol that represents a quantity, relationship, or other mathematical structure, but the essence of a variable is that it represents an entire set of quantities, relationships, or mathematical structures. When variables are introduced embedded in equations, students often develop the misconception that a variable stands for only one number—the value that makes the equation a true statement. Approaching the concept of a variable from a functional standpoint helps students understand how a variable can represent many different values. Variables are introduced as a natural extension of the numerical process. Consider the following situation:

> The second number is 3 times the first number, plus 2. Complete the table of values shown in figure 8.21. (The completed table is shown in fig. 8.22.)

First Number	Second Number
2	
5	
8	
13	
x	

Fig. 8.21. Blank table

First Number	Second Number
2	8
5	17
8	26
13	41
x	$3x + 2$

Fig. 8.22. Completed table

The introduction of the variable in the last line of the table is a natural extension of the numerical examples in the first lines of the table. There is no one "correct" value for the variable. Each line of the table expresses a value of the second number based on the value of the first number. Using the variable to express the generalized relationship $3x + 2$ represents every other possible line in the table.

Numerical problem solving as done in the previous examples in this article will develop students' understanding to the point where generalizing with variables is an obvious next step. When they are ready to make these generalizations, the multiline display can be used to reinforce the concept of a variable.

The **Replay** and editing functions of the graphing calculator can be used to efficiently analyze the situation discussed above. Figure 8.23 shows the first line in the table entered on the screen. Using the **Replay** function, repeat and edit the first value in the table, 2, to the second value, 5, and recalculate. Figure 8.24 shows the screen with the second line of the table calculated. Repeat this process for each line in the table (see fig. 8.25). Additional lines of the table can be calculated in the same manner by entering any chosen first number.

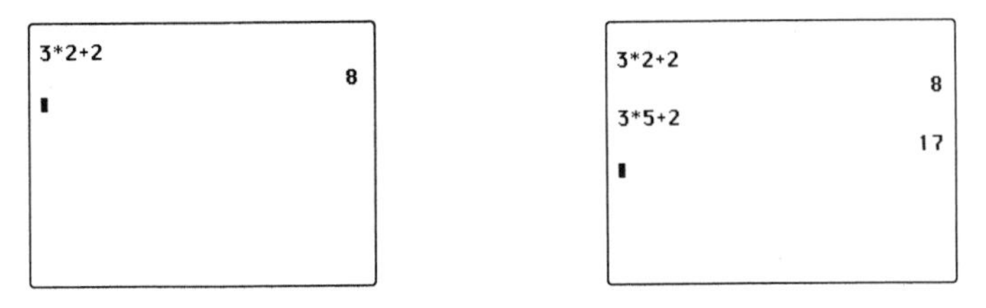

Fig. 8.23. First line of the table

Fig. 8.24. Second line of the table

3*5+2
17
3*8+2
26
3*13+2
41

Fig. 8.25. Screen showing the table values

In each step the only number in the mathematical phrase that was changed represented the values of the first number from the table. The screen provides a visual reference for the table-building process. The student always has several previous calculations to guide the next step.

At each step in the problem-solving process, the value of the first number is placed in the mathematical phrase and then that value is used to calculate the second number. This process emphasizes the idea that the first number can assume any value we wish and that the value of the second number is a function of the first number. This technique for generalizing a relationship can be used in many problem situations similar to the one presented here or in the previous examples.

MULTIPLE REPRESENTATIONS: BUILDING TABLES, PLOTTING POINTS, AND GRAPHING LINES

Although students often come to prealgebra or beginning algebra with some experience plotting points in a grid system, they usually have no formal training in the use of scale, in plotting points with integer or fractional coordinates, or in other conventions concerned with graphing mathematical relationships. The graphing calculator provides a unique environment for students to integrate the numerical, graphical, and symbolic representations of mathematical relationships. The integration of these three representations in an immediate, temporal framework helps students understand the relationship between numerical values in a table, the symbolic rule relating table values, and the corresponding graphical representation of the table and rule. Let us examine the following problem:

> The second number is half the first number plus 1. Complete the following table (fig. 8.26) and plot a set of points with the first number on the horizontal axis and the second number on the vertical axis.

The method using **Replay** and edit, described in the previous example, can be used to complete the table. Figure 8.27 shows the completed table of values. The last line in the completed table shows the general rule for finding the second number.

We shall use the completed table to set the **Range** of the graphics screen prior to plotting the pairs of values from the table. Set the horizontal axis from -10 to 10 with scale values (tick marks) of 1 and the vertical axis from -5 to 5 with scale values of 1. These settings define a viewing rectangle of $[-10, 10]$ by $[-5, 5]$.

Figure 8.28 shows the screen with the command to plot the first pair of values from the table, and figure 8.29 shows the result of this command on the graphics screen.

Return to the home screen (press CLEAR) and use the **Replay** function to repeat the last command line. Edit the values to reflect the next line of the table. Figure 8.30 shows the screen with several edited values (each command was edited and entered separately), and figure 8.31 shows the result of these commands on the graphics screen. Repeat this process for

First Number	Second Number
−8	
−6	
−4	
−2	
0	
2	
4	
6	
8	
x	

Fig. 8.26. Blank table

First Number	Second Number
−8	−3
−6	−2
−4	−1
−2	0
0	1
2	2
4	3
6	4
8	5
x	$.5x + 1$

Fig. 8.27. Completed table

each pair of values in the table. Students should notice that the points seem to be in a straight line; they are!

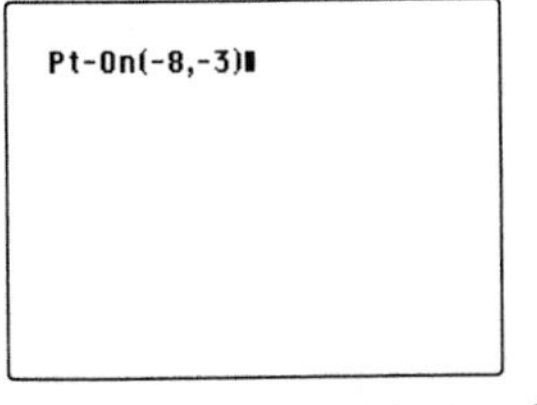

Fig. 8.28. Plot the point (−8, −3)

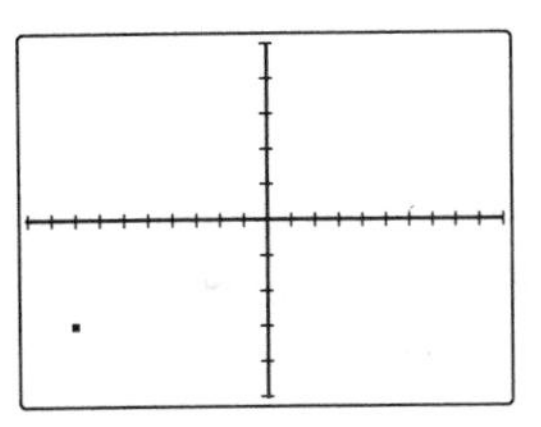

Fig. 8.29. Resulting graphics screen

```
Pt-On(-8,-3)
Pt-On(-6,-2)
Pt-On(-4,-1)
Pt-On(-2,0)
Pt-On(0,1)
Pt-On(2,2)
Pt-On(4,3)
```

Fig. 8.30. More points entered

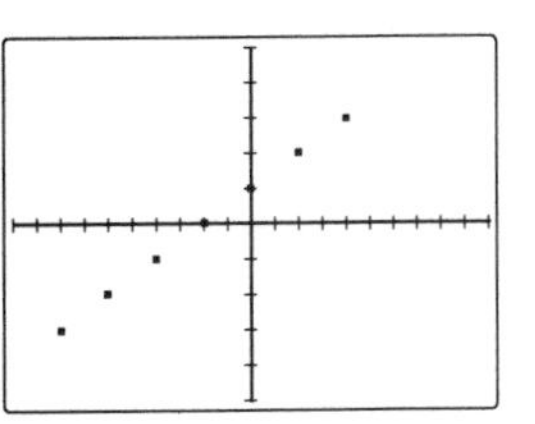

Fig. 8.31. Resulting graphics screen

As an extension of the table, have students add more points to the graph by choosing first values between those listed in the table (i.e., −1, 3, or 5). As new values are added to the table and plotted, the points always remain in the linear configuration. Students may add as many new points to the graph as they wish.

The final step in this process is to graph the general relationship $y = 0.5x + 1$ and show that this general graph contains all the points from the table and all other plotted points. The last line of figure 8.32 shows the command used to draw the graph, and figure 8.33 shows the screen after the graph has been drawn. The graph passes through all the plotted points.

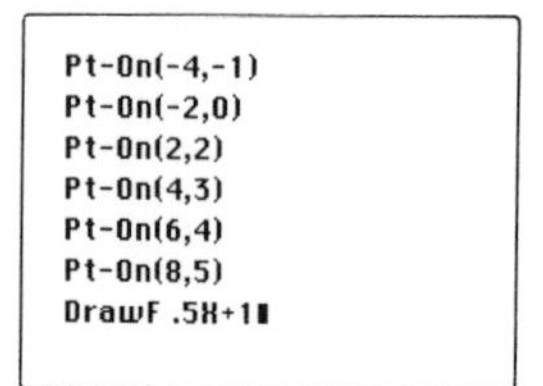

Fig. 8.32. Commands to draw the graph

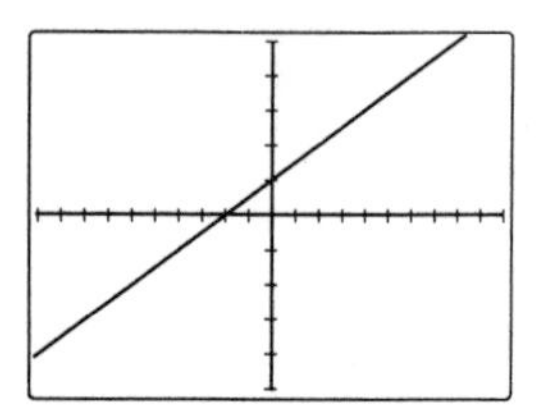

Fig. 8.33. Completed graph

With the graphing calculator available, students can gain valuable experience with graphs because they can reproduce many graphs in any scale very quickly. Through multiple, guided experiences with graphs, these students will begin to develop intuition about functions and their graphs and will begin to see the relationship between the algebraic representation and its graph.

CONCLUSION

Multiline display calculators furnish middle school students and teachers with a rich working environment for concept development, exploration, and numerical problem solving. A multiline display gives students a powerful learning tool to assist them in exploring and discovering numerical patterns and processes. The graphical capability of the graphing calculator also helps students make connections among the numerical, symbolic, and graphical representations of a mathematical relationship. The graphing calculator has the unique ability to integrate these multiple representations in an understandable way, giving teachers and students a powerful, portable, and yet affordable environment for teaching and learning mathematics in the middle grades.

REFERENCE

National Council of Teachers of Mathematics. *Curriculum and Evaluation Standards for School Mathematics*. Reston, Va.: The Council, 1989.

9

Calculators in the UCSMP Curriculum for Grades 7 and 8

Daniel B. Hirschhorn
Sharon Senk

DESPITE numerous recommendations to use calculators for instruction, by the end of the 1980s most middle school mathematics classes were not using calculators regularly (Kouba and Swafford 1989). In a 1985–86 survey, Weiss (1987) found that among mathematics teachers only about 38 percent in grades 7 to 9 thought their texts had good suggestions for the use of calculators. Because, as suggested by Begle (1973), the textbook has such a powerful influence on what students learn, one factor accounting for the low incidence of calculator use has been the lack of textbooks incorporating calculator usage into the core curriculum.

Since 1983, the University of Chicago School Mathematics Project (UCSMP) has been working to improve mathematics education in grades K–12 in the United States. Among its objectives is the incorporation of mathematically and pedagogically sound uses of calculator technology at all grade levels. At virtually every grade level calculators have had an impact on both the scope and the sequence of the course content.

For grades 7–12, the Secondary Component of UCSMP has designed and produced a set of textbooks for average students (whether college bound or not) and has conducted studies to determine how implementable and effective the materials are in typical school settings. The first two textbooks in the series are *Transition Mathematics* (Usiskin et al. 1990) and *Algebra* (McConnell et al. 1990), designed primarily for students in grades 7 and 8, respectively.

In this article we give specific examples of how calculators have changed both the scope and the sequence of the content designed to be covered in the middle school, describe the studies conducted by UCSMP to evaluate this curriculum, and summarize the results of these studies with respect to

We would like to acknowledge the Amoco Foundation and the Carnegie Corporation of New York for providing funding for the UCSMP research studies mentioned in this article.

implementation and mathematics achievement. Although the results reported here apply only to the calculator-based materials developed and tested by UCSMP, we are certain that there are other models for effective calculator-based curricula.

IMPACT OF CALCULATORS

Scope and Sequence of Calculator Keys

Scientific rather than four-function calculators are used in UCSMP secondary courses for three main reasons. First, realistic applications often demand the use of very large or very small numbers. However, most four-function calculators do not allow the user to enter a number with more than eight digits. For instance, when one attempts to enter the population of the world, the national debt of the United States, or the weight of an atom on most four-function calculators, an error message is displayed. But there is nothing mathematically wrong with those numbers! Second, scientific calculators usually employ the standard algebraic order of operations, whereas four-function calculators usually do not. Third, on scientific calculators there are keys for special constants and functions that can be used for mathematics in the middle school that are unavailable on four-function calculators.

Throughout the UCSMP curriculum the scientific calculator is used as one of several tools for doing mathematics (other tools include compass, computer software, graph paper, protractor, and ruler). Whenever a calculator key is introduced, explicit instruction is given on the use of that key, how the use of that key may be related to the use of other calculator keys, and how the use of the calculator key is related to mental or paper-and-pencil algorithms students already know.

In the first UCSMP secondary course, students learn the

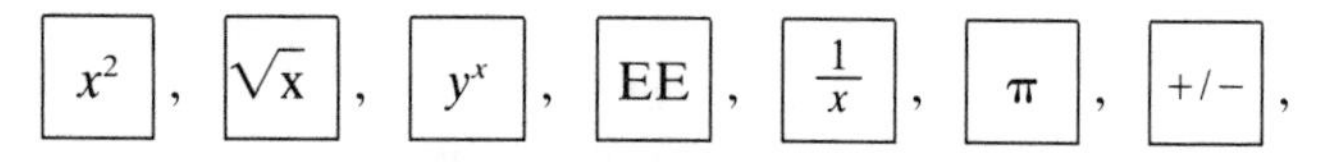

and memory keys while reviewing rational number concepts and procedures. In the following course, students learn the keys for factorial and constants while studying elementary concepts of counting, probability, and sequences.

Although occasionally there are questions that give students specific instructions about whether or not to use a calculator, typically the student is expected to decide what tool, if any, to use for a given problem. Students are expected to develop sound judgment to evaluate mentally expressions such as

$$\sqrt{100} \quad \text{or} \quad 2^3$$

and to approximate with a calculator expressions such as

$$\sqrt{479} \quad \text{or} \quad (3.2)^{16}.$$

Sample Content

The combination of scientific calculators and realistic applications allows students to be introduced to important mathematical ideas earlier and in richer contexts than is now typically done. The following sections illustrate how calculators assist in developing both traditional and nontraditional ideas related to one content strand—exponents—in the middle school years.

Transition Mathematics

Students are introduced early to positive integer powers and the use of the powering key (usually labeled $\boxed{y^x}$ on a scientific calculator). Using a calculator, students evaluate and compare powers of different bases, such as 2^{20} and 20^2. In the lesson following, they are introduced to scientific notation and possible displays on a calculator. Scientific notation allows students to solve realistic arithmetic problems. Included in the first assignment on scientific notation are questions such as the following (Usiskin et al. 1990):

- Rewrite 5,880,000,000,000,000,000,000 tons, the approximate mass of Earth, in scientific notation.
- If light travels about 1.60947×10^{10} miles per day, how many miles are in a light-year?
- How many seconds are there in a year?
- Why, when 531×10^{20} is entered into the calculator, is $\boxed{5.31\ 22}$ displayed?

Later, students learn about zero and negative integers as powers of 10 and their use in scientific notation for small numbers. They learn that 3.96×10^{-22} gram is an easier notation for the weight of one atom of uranium than 0.000 000 000 000 000 000 000 396 gram. With scientific calculators, such concepts can be grasped easily by average seventh graders. The use of powers and scientific notation throughout the rest of the course provides many experiences with exponents before algebra.

Algebra

Groundwork laid in the prealgebra concepts leads to a complete development of properties of exponents and an introduction to exponential growth. Consider the following example (McConnell et al. 1990):

You invest \$100 at an annual yield of 6%.(*a*) Graph your savings if you take the interest out of the bank and put it in a piggy bank. (*b*) On the same set of axes graph your savings if you leave the interest in the bank. Use values at the end of 0, 5, 10, 15, 20, 25, and 30 years.

By looking for patterns, first-year algebra students develop formulas for the value (V) of each investment at the end of year n:

$$\text{for part } a\text{: } V = 100 + 6n$$

$$\text{and for part } b\text{: } V = 100(1.06)^n$$

By using the formulas above, students find that at the end of 10 years under part *a* the amount is exactly $160, whereas under part *b* it is about $179. In computing these values, the scientific calculator achieves a significant savings in keystrokes over a four-function calculator. After 30 years, the amount under part *a* is $280, and under part *b* it is about $574.

With a scientific calculator, even students in first-year algebra can also begin to answer questions such as the following:

How many years will it take under each plan to double your money?

Notice that answering this question under part *a* involves solving the linear equation

$$100 + 6n = 200,$$

whereas answering the question under part *b* involves solving the exponential equation

$$100(1.06)^n = 200.$$

Students typically learn algorithms for solving linear equations in their first course in algebra and for solving exponential equations in their second algebra course. A scientific calculator opens up the discussion of approximation strategies such as guess and check long before students have learned traditional equation-solving algorithms. Alternatively, when studying such algorithms, the students can use the calculator as a check in evaluating expressions.

Having a scientific calculator also allows teachers and students to explore the meanings of exponents other than positive integers much earlier than usual. For example, average calculator-literate eighth graders will naturally ask about the meaning of expressions such as

$$100(1.06)^0 \quad \text{or} \quad 100(1.06)^{3.5} \quad \text{or} \quad 100(1.06)^{-2}.$$

The combination of the ability of the scientific calculator to evaluate such expressions and the ability of the student to check the reasonableness of the values in a familiar context helps pave the way for a formal look at all rational exponents in second-year algebra.

Obsolete Content

The middle school curriculum traditionally has been filled with content previously covered in the primary grades (Flanders 1987). Both because of calculators and because the students have been exposed to the content in earlier courses, there is no discussion of paper-and-pencil algorithms for whole number addition, subtraction, multiplication, or division in the UCSMP middle school curriculum. Operations with decimals are all performed on the calculator, and there is no discussion of the paper-and-pencil algorithms. It is assumed that the students are familiar with the paper-and-pencil algorithms associated with fractions, and although these skills are reviewed, they are not drilled. Applications are stressed, and there are applications that use decimals, fractions, mixed numbers, and integers. All the computations in an application are to be performed in the manner easiest for the student to carry out, whether it be mental, using paper and pencil, or with a calculator.

Similarly, the importance of applications and the availability of calculators have made some traditional algebra content obsolete. Most polynomials that arise from applications are not factorable over the rational numbers; and even those quadratic equations with integer coefficients are unlikely to be factorable. Thus, as a strategy for solving quadratics, UCSMP *Algebra* emphasizes the quadratic formula and de-emphasizes factoring techniques. Also, because expressions with radicals in the denominator are no harder to evaluate on a calculator than expressions with radicals in the numerator, there is very little work in algebra with what is often called *simplification of radical expressions*. (Factoring polynomials and rewriting radical expressions in order to recognize special patterns and to analyze functions are studied in the UCSMP courses that follow algebra.)

THE DESIGN OF EVALUATION STUDIES

As may be glimpsed from the previous discussion, all the UCSMP texts contain different approaches to traditional and nontraditional topics, often because of calculator technology. Since this was a significant change over other standard seventh- through twelfth-grade materials, evaluation studies investigated the effectiveness of this new curriculum. UCSMP evaluation studies examined two central questions: (1) To what extent are these curricula implementable in regular classrooms? and (2) what effects do these materials have on students' achievement in, and attitudes toward, mathematics?

About 2000 students from urban, suburban, and rural areas across the United States participated in each of the Transition Mathematics (TM) and Algebra field studies (Hedges et al. 1986; Mathison et al. 1989). The

UCSMP books used were preliminary versions (McConnell et al. 1987; Usiskin et al. 1985) of the texts currently available. In the TM study about 35 percent of the students were seventh graders, 45 percent eighth graders, and the rest in grade 9 or above. In the Algebra study, about 45 percent of the UCSMP students were eighth graders who would not ordinarily have been taking algebra at that time had they not had TM the year before; the other students were mostly average ninth graders, most of whom had not studied TM the previous year.

To ensure the equivalence of groups at the start of the school year, UCSMP and comparison classes from the same district were matched on the basis of their performance on a pretest covering standard prerequisite content. During the school year classes were observed and teachers were interviewed. At the end of the school year students were tested on both traditional and nontraditional content, and they were surveyed about their attitudes toward mathematics. (See Greene and Sabers [1967], Hanna and Orleans [1982], University of Chicago School Mathematics Project Secondary Component [1985], Wick and Gatta [1980], and Wick, Smith, and Gatta [1980] for the tests used.)

RESULTS

We report here on the implementation, attitude, and achievement measures as they relate specifically to calculator usage.

Implementation

In twenty-five of the TM classes calculators were purchased by the district or school; in the other nine classes students were responsible for providing their own calculators. In several of the latter, districts provided calculators for students who could not afford to buy them. Usually, students were allowed to take school-owned calculators home.

About 85 percent of TM teachers, but only 33 percent of comparison teachers, reported using calculators or other manipulatives daily. Among the TM teachers who used manipulatives virtually every day, more than 90 percent reported using calculators daily.

TM teachers reported that students tend to overuse the calculators at first but then settle in to appropriate usage after a couple of months. Similarly, a few teachers who were using school-owned calculators reported some calculator management problems at first, but these were solved early in the year.

In the Algebra field study, a marked difference toward the implementation of calculators in UCSMP and comparison classes was also noted. Table 9.1 shows the percent of teachers reporting various uses of calculators. All the

UCSMP classes had access to scientific calculators, all but one UCSMP *Algebra* teacher allowed students to use them whenever and however they wished, and all the UCSMP teachers allowed them on tests. In contrast, 46 percent of comparison teachers did not allow the use of calculators at any time, and the comparison teachers who allowed calculators were more restrictive. As can be seen by comparing the second column of table 9.1 with the last two columns, the comparison algebra teachers were typical of grades 7–12 mathematics teachers surveyed that year (Weiss 1987).

Table 9.1
Calculator Usage in Mathematics Classes (Percent of Teachers)

	Algebra		National Survey	
Type of Use	UCSMP	Comparison	7–9	10–12
Checking answers	96	45	23	29
Doing computations	100	54	27	47
Solving problems	96	33	22	37
Taking tests	100	22	10	35
Sample *N*	27	18	671	565

More than 80 percent of UCSMP *Algebra* teachers thought their texts had good suggestions for using calculators, in contrast to 15 percent of comparison teachers and 38 percent of grade 7 to 9 mathematics teachers in the Weiss (1987) survey.

Teachers of both UCSMP courses reported a clear difference between using a *scientific* calculator and using a *four-function* calculator. They noted that students expressed curiosity about the keys that they did not use in the course, such as those pertaining to logarithmic or trigonometric functions. Since the keys implied that there was more mathematics for students to learn after the completion of their current course, the scientific calculator was considered to be a good motivator for continuing to study more mathematics.

Student Attitudes

In the TM study, student attitudes were surveyed at the beginning and the end of the school year, but in the Algebra study, only at the end of the school year. Responses to the two calculator-related items are given in table 9.2.

The data do not give a clear answer on how student attitudes are affected by using the calculator. On the one hand, there is a strong perception by both UCSMP and comparison students that the use of calculators leads one to forget how to do arithmetic. On the other hand, over half of the UCSMP students at the end of the year believed that a calculator helped them to learn more mathematics.

Table 9.2
Student Attitudes toward Calculators (Percent Who Agree or Strongly Agree)

Item		TM	Comp	Alg	Comp
If you use a calculator too much, you forget how to do arithmetic.	Fall	43	58	—	—
	Spring	44	44	41	57
Using a calculator helps me learn math.	Fall	39	28	—	—
	Spring	51	39	58	22

Achievement

Results on the end-of-the-year achievement tests for all matched pairs of classes in the TM study are shown in table 9.3.

Table 9.3
Achievement in Transition Mathematics Study (Mean and Standard Error)

	TM		Comparison	
Test	$\bar{x}$	s.e.	$\bar{x}$	s.e.
General Mathematics				
without calculators	23.76	1.35	24.20	1.10
with calculators*	29.27	1.04	26.29	1.08
Orleans-Hanna Algebra Prognosis*	38.35	1.86	36.30	1.78
Geometry Readiness*	9.68	0.67	8.50	0.70

*Difference (TM-Comparison) is significant, $p < .05$; $n = 20$ throughout.

When a calculator was not allowed, the difference between the achievement of TM and comparison students on the General Mathematics Test was not statistically significant, but when a calculator was allowed, the TM students outperformed their peers. On both an algebra readiness and a geometry readiness test, the TM students scored significantly higher than the comparison students.

On an additional test covering the use of calculators, TM students also outperformed their peers. Figure 9.1 lists the six items from this test on which the largest differences in performance were found.

In the Algebra study on a standardized algebra test given at the end of the school year (on which no calculators were allowed), UCSMP and comparison students did equally well. Among the students in eleven matched pairs of classes, the mean achievement of UCSMP students was 51.5 percent, and the percent correct among comparison students was 50.3. Additionally, UCSMP *Algebra* students outperformed comparison students on UCSMP-designed tests, on which calculators were allowed, covering both traditional and nontraditional content.

Figure 9.2 shows the performance of both UCSMP and comparison students on items from both the standardized test (1–40) and the UCSMP-

1. $3/23$ is closest to
   (a) 0.13
   (b) 1.3
   (c) 0.77
   (d) 7.7
2. Which is another way of expressing 7.84×10^{-3}?
   (a) 7840
   (b) −235.2
   (c) 75.4
   (d) .00784
3. If you try to do −1 divided by 0 on a calculator, there will be an error message because:
   (a) Calculators cannot operate with negative numbers.
   (b) Calculators can do some operations with negatives, but cannot divide with them.
   (c) It is impossible to divide by 0.
   (d) The calculator is broken.
4. Which is another way of expressing 3.6×10^2?
   a) 36
   (b) 360
   (c) 3600
   (d) 36,000
5. 113π is closest to
   (a) 339
   (b) 350
   (c) 355
   (d) 1133
6. $920 + 921(922 + 923 \times 924)$ is about
   (a) 7.86
   (b) 7.86×10^8
   (c) 7.86×10^{10}
   (d) 7.86×10^{12}

Fig. 9.1. Calculator items on which *Transition Mathematics* students outperformed comparison students

designed test (50–140) on two content strands: percents and exponents. The area between the dotted lines highlights items where the difference between UCSMP and comparison students was within 10 percentage points. The area above the dotted lines highlights items where the UCSMP students outperformed comparison students by more than 10 percentage points. This figure indicates that, in general, work with calculators in these content areas benefited students.

On a subtest concerning technology, UCSMP *Algebra* students also outperformed comparison students on items testing their knowledge of how to use and interpret both a scientific calculator and simple BASIC programs.

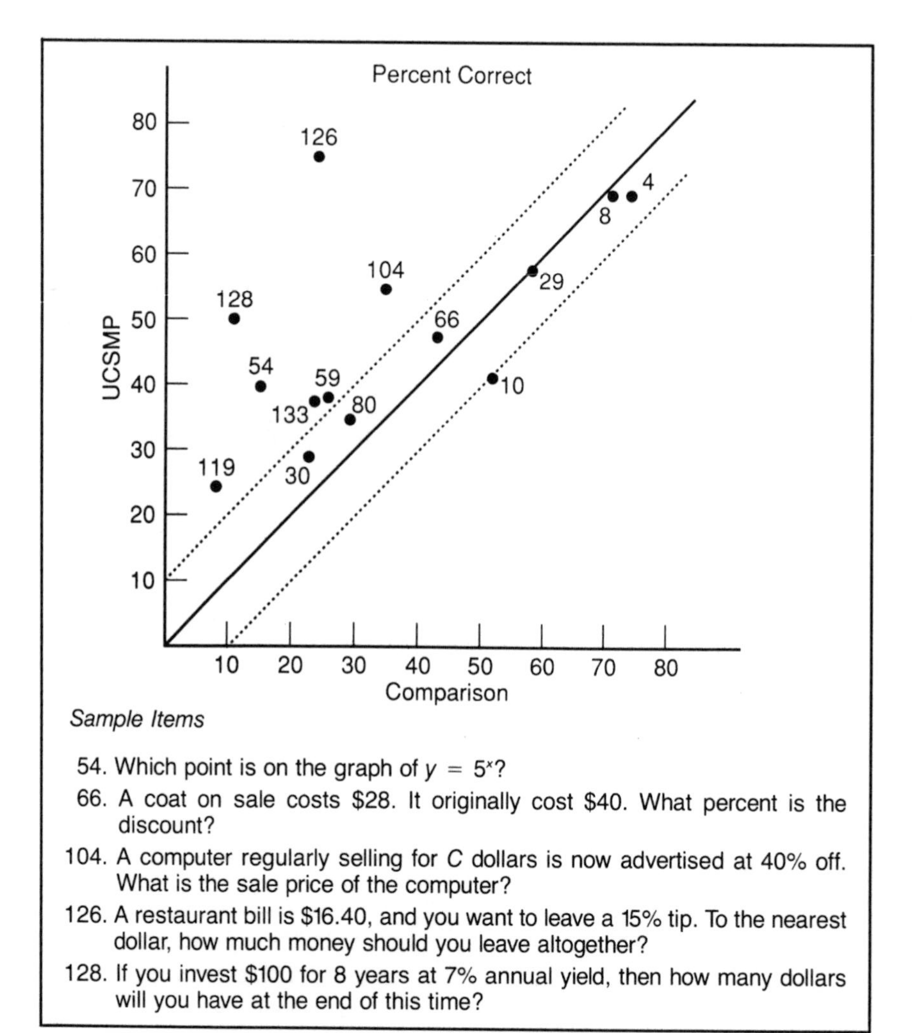

Fig. 9.2. Performance of UCSMP and Comparison Algebra students on items about percent, exponents, or exponential growth.

DISCUSSION

When the Transition Mathematics study was conducted in 1985–86, there was no major consensus in favor of using scientific calculator technology in the middle school curriculum. In fact, it was not clear that a curriculum that required the use of scientific calculators could be accepted by average middle school teachers or students. Such common technologies as the graphing calculator, powerful personal computers, and fax machines were in their infancy. There was still a great caution, if not reluctance, in the mathematics education community toward a curriculum that mandated the use of technology of any sort.

The research described in this article points out that teachers in grades 7 to 9 are able to implement the UCSMP scientific calculator–based curriculum in normal classrooms around the United States. Consistent with other studies, students who learn with the assistance of a calculator generally suffer no loss of standard arithmetic or algebraic skills. Furthermore, there are clearly documented benefits to mathematics achievement from this particular calculator-rich curriculum. First, when students are taught to use a scientific calculator and allowed to use it freely, they learn to use the calculator more effectively than those without such opportunities. Second, achievement improves on some standard content (e.g., applications of percents). Third, students are able to learn content that was formerly reserved for older or more able students. For instance, with the aid of a scientific calculator, average eighth graders are able to study exponential growth and compare and contrast it with linear growth.

The effects of calculators on students' attitudes are less clear. Keys not yet studied on a scientific calculator remind students that there is more mathematics to come in their secondary mathematics courses, thus serving as motivation to continue to study mathematics in high school. Students who use a calculator for the entire year will likely believe that a calculator helps them learn mathematics, but those who initially perceive that the use of a calculator leads to a deterioration of arithmetic skills will most likely not be swayed from this belief after a single mathematics course that uses a calculator.

As shown in this article and in Usiskin (1988), the everyday use of scientific calculators in the middle school curriculum makes some content obsolete, makes other content more important, and changes the ways still other content is viewed. By writing materials that incorporate the use of calculator technology into the core secondary school curriculum, the University of Chicago School Mathematics Project has demonstrated that textbooks that make these content changes can exist. Whereas this proves the existence of such materials, we hope they will not be unique. We encourage further development and research in this area.

REFERENCES

Begle, Edward G. "Some Lessons Learned by SMSG." *Mathematics Teacher* 66 (March 1973): 207–14.

Flanders, James R. "How Much of the Content in Mathematics Textbooks Is New?" *Arithmetic Teacher* 35 (September 1987): 18–23.

Greene, Harry A., and Darrell Sabers. *Iowa Achievement Aptitude Test.* 3d ed. Iowa City: Division of Continuing Education, University of Iowa, 1967.

Hanna, Gerald S., and Joseph B. Orleans. *Orleans-Hanna Algebra Prognosis Test.* San Antonio, Tex.: The Psychological Corporation, Harcourt Brace Jovanovich, 1982.

Hedges, Larry V., Susan S. Stodolsky, Sandra Mathison, and Penelope V. Flores. *Transition Mathematics Field Study.* Chicago: University of Chicago School Mathematics Project, 1986.

Kouba, Vicky L., and Jane O. Swafford. "Calculators." In *Results from the Fourth Mathematics Assessment of the National Assessment of Educational Progress,* edited by Mary M. Lindquist, pp. 94–105. Reston, Va.: National Council of Teachers of Mathematics, 1989.

McConnell, John W., Susan Brown, Susan Eddins, Margaret Hackworth, Leroy Sachs, Ernest Woodward, James Flanders, Daniel Hirschhorn, Cathy Hynes, Lydia Polonsky, and Zalman Usiskin. *Algebra.* Field trial ed. Chicago: University of Chicago School Mathematics Project, 1987.

———. *Algebra.* Glenview, Ill.: Scott, Foresman, 1990.

Mathison, Sandra, Larry V. Hedges, Susan S. Stodolsky, Penelope V. Flores, and Catherine Sarther. *Algebra Field Study.* Chicago: University of Chicago School Mathematics Project, 1989.

University of Chicago School Mathematics Project Secondary Component. *Geometry Readiness Test.* Chicago: University of Chicago, Department of Education, 1985.

Usiskin, Zalman. "The Beliefs Underlying UCSMP." *UCSMP Newsletter* 2 (Winter 1988): 9–15.

Usiskin, Zalman, James Flanders, Cathy Hynes, Lydia Polonsky, Susan Porter, and Steven Viktora. *Transition Mathematics.* Field trial ed. Chicago: University of Chicago School Mathematics Project, 1985.

———. *Transition Mathematics.* Glenview, Ill.: Scott, Foresman, 1990.

Weiss, Iris R. *Report of the 1985–86 National Survey of Science and Mathematics Education.* RTI/2938/00-FR. Research Triangle Park, N.C.: Research Triangle Institute, 1987.

Wick, John W., and Louis A. Gatta. *High-School Subject Tests: Algebra, Form B.* Glenview, Ill.: Scott, Foresman, 1980.

Wick, John W., Jeffrey K. Smith, and Louis A. Gatta. *High-School Subject Tests: General Mathematics.* Iowa City, Iowa: American Testronics, 1980.

10

Personal Technology and Classroom Change: A British Perspective

Kenneth Ruthven

Part 3

IN RECENT years, a new generation of electronic devices has emerged, described both as advanced calculators and as pocket computers. Resembling scientific calculators, these machines offer facilities previously available only on traditional computers, notably environments for mathematical graphing and programming.

My classroom experience with advanced calculators has been as a teacher of upper-level secondary school mathematics and as coordinator of the Graphic Calculators in Mathematics development project. In this project six small groups of teachers worked with classes of high school students who had permanent access to such calculators throughout their advanced mathematics course. Rather than following any prescribed program of calculator activities, the teachers were free to plan the work of their classes within the normal constraints of the syllabus, meeting periodically to exchange ideas and review progress.

Five of the groups worked with the Casio fx-7000G model, and one used the Hewlett-Packard 28C. These were the principal models available in Britain at the start of the project; whereas the Casio machine was considerably cheaper, the Hewlett-Packard offered more facilities, notably symbolic manipulation.

Although there are important differences between the British and North American contexts, I shall focus here on the fundamental issues that I believe to be common to both, in the hope that my insights will be of some general interest and value.

THE CALCULATOR AS AN INDIVIDUAL RESOURCE

In Britain, the potential of computing in teaching and learning mathematics has been widely acknowledged but little realized. A major impediment has been limited access to computing facilities. The occasional mathematics lesson in a central computer room encourages the view of computing as an exceptional activity, isolated from the mainstream of mathematics. The goal of many schools now is to equip each mathematics classroom with a single microcomputer, so that computing may become a more integral part of normal mathematical activity. But the classroom microcomputer is often used predominantly for demonstration by the teacher, and even where direct use by students is encouraged, access is usually limited and under the teacher's control. In these circumstances, few students are able to make spontaneous use of computing facilities.

Advanced calculators now offer access to computing facilities at a fraction of the previous cost; a class set of calculators is comparable in expense to a single classroom microcomputer. Equally, increasing numbers of students are acquiring their own advanced calculator rather than a scientific calculator. Against this background, the Graphic Calculators in Mathematics project treated the advanced calculator as an *individual resource,* readily available to each student not only in mathematics lessons but during lessons in other subjects and private study. In some project schools, each participating student had a machine on permanent loan; in others, calculators were freely available in every mathematics lesson and from the school library or resource center at other times.

The impact of this unrestricted access was impressive. As with any new technology, the initial period of becoming familiar with the calculator could be frustrating at times, but knowing that the resource would always be available was a powerful incentive to learn to use it. This learning could take place more privately and informally than on a classroom microcomputer with its public display exposed to scrutiny by hovering enthusiasts all too ready to offer advice or take control.

After one school term, nearly all the project students were making confident and spontaneous use of the calculating and graphing facilities of the advanced calculator. In particular, virtually all those with a machine on permanent loan used it in preference to their previous calculator. For a small number of students, however, a lack of confidence in operating the advanced calculator persisted. Although these students did become proficient in using the distinctive graphing facility, they preferred to retain the familiar procedures of their previous machine for calculating. An important factor here was the support and encouragement offered to students; in the two project classes where the teacher showed strong reservations about using the calculator, disproportionate numbers of such students were found.

Still smaller in number were the students who were reluctant to use a calculator of any type because they felt that by doing so they "lost control" of the mathematics. They wanted to take nothing as given, retaining direct responsibility for, and detailed awareness of, every part of a mathematical process. Although this is an admirable expression of intellectual fastidiousness, the students themselves, particularly as they met more complex mathematical situations, clearly recognized its drawbacks: likelihood of error, loss of time, and distraction from the central argument. In practice, such students made increasing use of a calculator, and their reluctance to relinquish control encouraged them to interpret results particularly critically.

Unlike calculating and graphing, programming is not integral to advanced secondary school mathematics in its present form. Consequently, this aspect of calculator use was not given equal stress by all the project teachers. However, after one school year, around two-thirds of the students were making confident and spontaneous use of the programming facility. Here, some gender differences were noted: in particular, a lower proportion of female students expressed confidence in programming, although this must be judged in the light both of the tendency of male students to be less self-critical and of the lower proportion of female students with prior programming experience. Typically, student-devised programs were short and specific to a particular problem, often consisting simply of a sequence of commands that needed to be carried out several times. In addition, many students kept a few programs of more general application permanently stored; the most popular types scaled the graphing axes in some way or solved quadratic equations.

STUDENT STRATEGIES AND THE CALCULATOR

Students tended to use the calculating facilities of the advanced calculator as they had those of the scientific calculator, simply as a convenient substitute for mental and written methods that would be unduly complex and time-consuming. Similarly, their use of the graphing facilities quickly replaced extended mental and written routines such as graphing an expression by calculating and plotting points. However, other than as the result of deliberate teaching, the underlying mathematical approaches employed by students were usually no different from before. In this sense, their use of the calculator tended to be *traditional*. Nonetheless, as confidence grew, there were more examples of *innovative* calculator use by students. A good illustration arose from the following word problem on sequences:

> A sum of \$1000 is invested in an account in which interest of 1% is added at the end of each month. After how many months will the sum in the account exceed \$1400?

The approach expected by the teacher was to formulate the question symbolically, as $1000(1.01)^n > 1400$, for example, and then to manipulate the inequation progressively to arrive at a form allowing direct evaluation, such as $n > \log(1.4)/\log(1.01)$. One student response, however, was simply to enter 1000 on the calculator and then repeatedly multiply by 1.01 until the displayed value exceeded 1400, maintaining a count of the number of multiplications—a *building-up* strategy. A more sophisticated response involved repeatedly comparing the value of an expression of the form $1000(1.01)^n$ with 1400, starting from a rough guess and progressively improving it—a *trial-and-improve* strategy.

Another illustration occurred when I introduced advanced calculators to a class for the first time, as a convenient means of drawing graphs during some work on polynomials. Later, I set the following task to give students an opportunity to make use of polynomial expansions:

> The equation $4x^3 - 3x^2 - 10x - 49 = 0$ has a solution that is approximately 3. By substituting $x = 3 + h$ and ignoring squares and higher powers of h, find a closer approximation to the value of the solution. Write a program to improve on this solution.

I had added the last sentence knowing that several of the students were already proficient microcomputer programmers and thinking that they might devise a numeric iteration. I had not expected graphic iteration! Ignoring my first suggestion, one group graphed the polynomial over the range from 3 to 4, marking intervals of 0.1 on the axis, but they found no solution. Next, they graphed from 2 to 3 and found that the solution lay between 2.9 and 3. They then graphed over this range, reducing the axis intervals to 0.01, and read off a still narrower range. In this way, they proceeded to find ever closer bounds for the solution.

THE CALCULATOR AS A COGNITIVE TOOL

Although symbolic manipulation is very much a part of current advanced mathematics courses, the facilities available on one of the calculator models were little used, largely because of a lack of correspondence between the way in which students thought about manipulation and the manipulative operations available on the machine. For example, to achieve what students saw as a simple matter of rearranging the order of terms called for a demanding combination of associative and commutative transformations (fig. 10.1). Thus, there was a considerable mismatch between the informal concepts of students and the formal "language" of the calculator, and students found calculator procedures cumbersome and slow. This points to the complex relation between mathematical thinking and calculator use—the nature of the calculator as a *cognitive tool*. Essentially, the calculator operations for symbolic manipulation reflected a sophisticated model of algebraic structure that could not be easily assimilated to the more informal conceptions of students. In practice, the students (and their teachers) made a pragmatic

```
'(n+3)^2'
EXPAN
'n^2+2*n*3+3^2'
COLCT
'9+n^2+6*n'
FORM
((9+(n^2))+(6*n))
[→]...[→]
((9+(n^2))+(6*n))
A→
(9+((n^2)+(6*n)))
↔
(((n^2)+(6*n))+9)
ON
'n^2+6*n+9'
```

Fig. 10.1. Expanding $(n + 3)^2$ on an advanced calculator

judgment that the costs of the conceptual change required to make use of the facility exceeded its benefits.

A similar issue arose with the encoding of mathematical expressions. At first, users of each model of advanced calculator had some difficulty in breaking keystroke-sequencing habits established on conventional machines. It was understandably much easier to adapt to the (almost) standard mathematical notation used by Casio—an approach that might be described as "what you write is what you key"—than to the Reverse Polish Notation adopted by Hewlett-Packard. Indeed, once adaptation had taken place, the use of standard notation seemed to reduce cognitive load by removing the need to translate between languages or to formulate an expression in a language specific to the calculator.

COGNITIVE GROWTH AND THE CALCULATOR

Ideally, a cognitive tool should not only be capable of assimilation to established modes of thinking but also be able to support cognitive growth and change on the part of the user. Indeed, this is an important element of the rationale for using calculators in the mathematics classroom: that they offer not simply a mechanism for calculating and drawing but a medium for thinking and learning.

Evidence from a small research study (Ruthven 1990) into the influence of the graphing facility illustrates the way in which the use of a calculator can promote mathematical development. In the study, the performance of project students was compared with that of similar students without access to advanced calculators (or computer graphing). Toward the end of the first year of their courses, students were tested on the reverse task to that automated by the calculator, that of describing a given graph in symbolic terms (fig. 10.2). Tested on six such items, the attainment of the project group was substantially and significantly higher than that of the comparison group. There was also a significantly differential influence on the attainment of male and female students: in the project group, female students outperformed male; the reverse pattern was found in the comparison group.

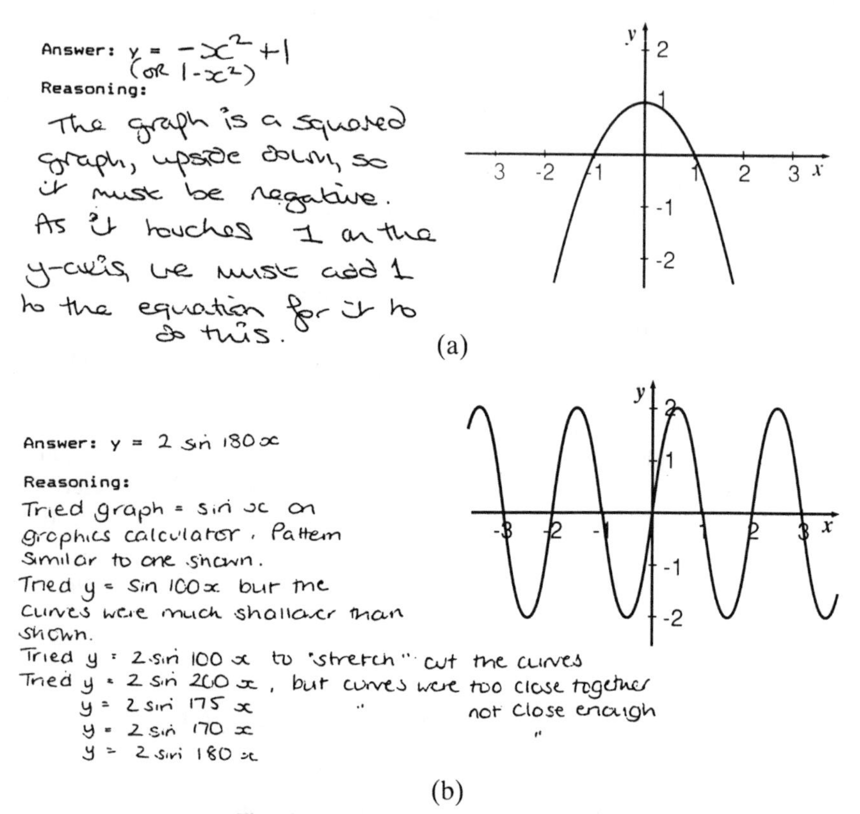

Fig. 10.2. Two student responses to test items

The performance of the project group was superior in two respects. First, they were better at simply recognizing a graph as being of a particular type, such as quadratic or sine. Next, they were more successful in building up a precise symbolic description of the graph by exploiting salient information—such as its orientation and the position of its extreme values, zeroes, and asymptotes—and their knowledge of the relationships between these features of a graph and its symbolization. Although most members of the project group used a conventional approach based on direct analysis of the graph (fig. 10.2a), there was also some use of a trial-and-improve approach (fig. 10.2b).

When students start to use a calculator, they see it as simply offering a limited range of *automatic* procedures for such familiar processes as evaluating an expression or drawing a graph. Growing confidence, however, can encourage novel approaches, most prominently variations on trial-and-improve. Although the trial-and-improve approach has limitations, it often gives students a means of tackling a problem that would otherwise be intractable. Moreover, it is an example of a *moderated* procedure, dependent for its efficiency and success on intermediate judgments made by the user. As the user becomes increasingly familiar with a particular type of mathematical situation, such judgments tend to become more sophisticated. In the example above, the student is not guessing blindly but is interpreting graphic feedback in the light of key mathematical principles. Indeed, in the research study, there was evidence that the use of a trial-and-improve approach can help to trigger the critical insight needed to switch to a more direct analytic approach.

DEVELOPING TEACHING WITH THE CALCULATOR

Much educational innovation fails because it ignores the mediating role of teachers. The Graphic Calculators in Mathematics project took the view that real educational innovation, although drawing on the ideas and experience of others, actually takes place at the classroom level and depends on thoughtful change on the part of teachers. The project was organized around the professional development of six small groups of teachers nominated by their school districts. These teachers came to the project with a variety of experience and attitudes. None had met an advanced calculator; few were making regular use of computers in their teaching; for most, scientific calculators were simply a marginal tool in the classroom. Some were enthusiastic about the prospect of working with advanced calculators, most were open-minded, a few were positively sceptical.

We met twice a year for a three-day conference intended to encourage reflection on professional practice and the exchange of ideas and experiences. Sometimes there were short, plenary presentations by participants on approaches to standard topics, on situations lending themselves to calculator exploration, or on work done by students. Much of the activity, however, took place in smaller groups: describing and discussing classroom activities that members had initiated or devising teaching approaches and materials. Between conferences, each local group met occasionally, using similar methods, but with one valuable addition: visits to the classrooms of participating teachers either to observe or take joint responsibility for a lesson, followed by discussion.

For me, the project has highlighted two central issues of teaching methodology: the role of algebraic symbolism in developing mathematical ideas and the locus of intellectual authority in the classroom. Many teachers came to the project most confident with mathematical approaches based on algebraic symbolization and manipulation and with classroom methods that emphasized the teacher's control of student learning.

ALTERNATIVE MATHEMATICAL APPROACHES AND THE CALCULATOR

Traditionally, algebraic symbolism has been the primary medium for developing ideas in advanced mathematics. This reflects its power as a means not only of summarizing patterns and relations in diverse situations but of reformulating them by manipulating their symbolic representations. Given concern that many students lack confidence and competence in algebra, the advent of calculators with facilities for symbolic manipulation has been hailed as a breakthrough. My view is rather different. Earlier, I suggested that there may be a substantial mismatch between students' informal conceptions and the manipulative environment provided by the calculator. Under these circumstances, the use of a calculator may actually increase cognitive demands.

Moreover, I have reservations about a teaching approach that rests too heavily on the symbolic medium, for such an approach tends to detach a mathematical argument from its original context. Our focus becomes the symbols as objects in themselves rather than as representations of the elements of some underlying situation. By manipulating these symbols, we reformulate this situation indirectly, rather than directly reorganizing our conceptualization of it. Often, students' difficulties have as much to do with making sense of the representational function of symbols as with manipulating them. Indeed, most courses pay insufficient attention to issues of representation, treating manipulation as an end in itself rather than as a means to reformulating mathematical relations.

The power of the calculator lies in facilitating classroom approaches in which mathematical relations are first explored through numeric or graphic representations of particular instances. As an illustration, consider the standard topic often described as $a \cos x + b \sin x$ in the form $r \cos(x - t)$. The conventional teaching approach simply assumes that such a relationship exists, reducing the problem to a matter of deriving r and t from a and b by a process of symbolic manipulation. This usually has to be demonstrated by the teacher; not only do many students find the manipulation hard, but they are not helped by the lack of any developed sense of the relation behind the symbols.

My experience is that a more powerful starting point is for students to use a graphing facility to explore what happens when sine and cosine expressions are combined to give expressions such as $\cos x + \sin x$, $\cos 2x +$

sin x, and 2 cos x + sin x (fig. 10.3), looking for alternative descriptions for each graph and for any pattern in the results as a whole. Such an approach depends, of course, on the familiarity with transformational ideas and with the corresponding relationships between symbolic and graphic forms that was characteristic of the project students. Indeed, this would seem to be a key strand in any curriculum aiming to exploit the graphing capabilities of advanced calculators.

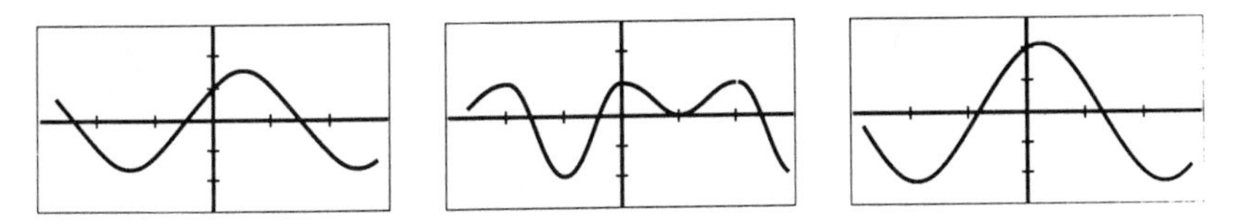

Fig. 10.3. Combinations of sine and cosine expressions

Two conjectures usually emerge: (1) that there are simple alternative descriptions for certain graphs as a single sine or cosine expression, and (2) that the existence of such descriptions depends on whether or not the "angles" of the two components are the same. The commonest formulations of the graph of cos x + sin x, for example, are as a cosine graph shifted a little to the right or as a sine graph shifted a little to the left. By moving the screen cursor of the calculator, students can determine the coordinates of critical points on the graph—for example, that one maximum is (45, 1.4)—and thus arrive at a more precise description, such as 1.4 cos(x − 45) or 1.4 sin(x + 45), which can be tested directly by superimposing it on the original graph. Recourse to familiar formulas for cos(a − b) and sin(a + b) as a means of explaining these alternative descriptions is, if not always spontaneous, easy to encourage. Symbolic manipulation now clarifies the relation and reveals that 1.4 is simply an approximation to $\sqrt{2}$, something that can be confirmed by evaluating cos 45 + sin 45.

The advantage of this mathematical approach is that it brings out underlying aspects that are often glossed over, giving rise to misconceptions. What is the relation about? Under what conditions does it hold? Why does it hold? The initial examination of a range of particular examples, treated graphically, helps students to formulate the nature of the relation more clearly and fully; it informs the search for an appropriate generalization; and it can facilitate a later transition to an essentially symbolic treatment. Such an approach is particularly helpful in making ideas more accessible to students who would struggle with an approach dependent on algebraic methods from the very start.

NEW CLASSROOM ROLES AND THE CALCULATOR

One advantage of this classroom method is that responsibility is devolved to students so that they play a more active part in developing and evaluating mathematical ideas. This can help students not only to grasp new ideas but also to develop the capacity to tackle novel situations. Rather than depending on the teacher to approve their ideas, students use the calculator to evaluate these ideas for themselves. This also means that the acceptability of a piece of mathematics is seen to depend not on the authority of the teacher but on its being consistent with some accepted wider body of mathematics. Here, the role of the teacher is to create a mathematical situation from which important concepts and relations are likely to emerge and, through sensitive intervention, to support students in exploring this situation and clarifying their ideas.

Let me conclude with a representative comment from a project teacher:

> I was initially fairly sceptical of putting graphic calculators to use in teaching mathematics. However, through involvement in the project, I have been enlightened. I am continually impressed by the ease with which students learn to use and master the machine; in fact, they are forever making new discoveries and devising ingenious strategies for solving problems. It is fair to say that the introduction of the graphic calculator has revolutionized my approach to the teaching of many mathematical topics, and I have passed on this enthusiasm and knowledge to my colleagues, who are now all converted.

REFERENCE

Ruthven, Kenneth. "The Influence of Graphic Calculator Use on Translation from Graphic to Symbolic Forms." *Educational Studies in Mathematics* 21 (1990): 431–50.

11

Teaching the Line of Best Fit with a Graphing Calculator

Rheta Rubenstein

THE *Curriculum and Evaluation Standards for School Mathematics* (NCTM 1989) challenges us to change our vision of what mathematics is and how it can best be learned. In particular, the algebra and statistics standards for grades 9–12 advocate that our curricula integrate concepts and applications that help students relate equations and data, fit curves to data, make interpretations and predictions from data, and, for college-intending students, understand and use techniques of data transformation. Further, the *Standards* suggests classroom practices that are centered on discussion and investigation and that make mathematical ideas accessible to a wider range of students. One linchpin in this vision is the availability of statistical-graphing calculators capable of producing scatter plots, graphs and parameters for lines or curves of best fit, and correlation coefficients. This article describes one illustration of the document's vision—a unit of instruction I have used at the high school level to develop and apply the line-of-best-fit concept using graphing-statistical calculators.

As with any technology-enhanced instruction, this unit involves a number of challenges:

- Paving the way so that students can make sense of the results the calculator produces
- Balancing appropriate calculator use with noncalculator experiences that develop concepts
- Impressing on students the need to interpret calculator outputs
- Identifying and sequencing interesting and appropriate applications

The instructional sequence gives an example of how these concerns and the recommendations of the *Standards* can be addressed.

PRELIMINARY ACTIVITIES: DATA COLLECTION, BY-HAND GRAPHING, AND INTERPRETATION

The unit begins by having students collect and organize data about themselves: weight, height, armspan, handspan, the distance from the ceiling to the fingertip of their raised arm, and the value of their pocket change. By design, some of the variables are chosen to have strong correlation (height and armspan); weaker, but positive correlation (height and weight); negative correlation (distance to ceiling and height); and zero or very low correlation (pocket change and any of the others).

Before more formal methods are used, it is helpful to ask students to use their natural intuition about the variability of human measures and discuss some questions:

- Do you think tall people will have generally longer armspans?
- Suppose two people have nearly the same armspan. Does this mean they must have nearly the same height?
- What sort of relationship do you expect there to be between a person's handspan and the amount of change in their pocket?
- What relationship do you expect between a person's height and the distance from their fingertip to the ceiling?

Next, students produce, by hand, a scatter plot for the height-weight data. A sample is shown in figure 11.1. This activity raises the issues of selecting an appropriate viewing window and scaling units for the axes. Students discuss general trends, variability, and, if they appear, outliers. Students can then be guided through fitting a line to the data by hand. Next, the following issues (fig. 11.2) are discussed to integrate and extend earlier knowledge from algebra.

Person	Height (cm)	Weight (kg)	Person	Height (cm)	Weight (kg)
1	162	50	13	168	52
2	155	49	14	164	65
3	172	68	15	163	55
4	163	45	16	164	52
5	163	50	17	160	50
6	166	49	18	170	53
7	168	58	19	163	57
8	170	61	20	159	50
9	159	52	21	168	58
10	155	47	22	168	67
11	172	61	23	157	47
12	169	55	24	163	47

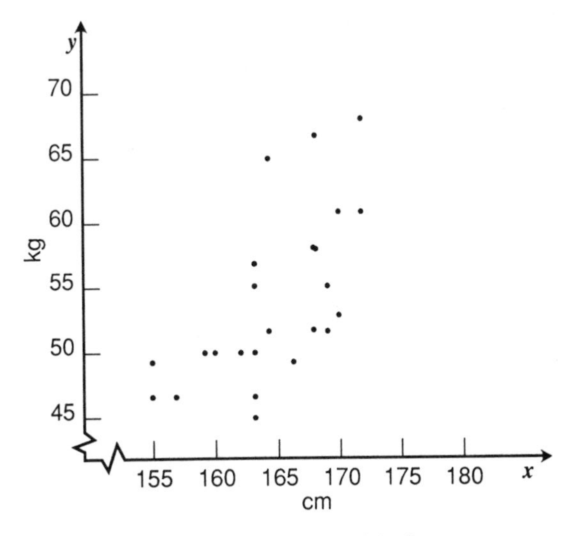

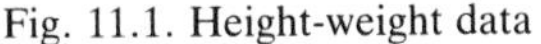

Fig. 11.1. Height-weight data

Follow-Up Questions to By-Hand Graphing

1. Find the slope of your line.
2. What does the slope represent in terms of the original data?
3. Find the y-intercept of your line.
4. What does the y-intercept represent in terms of the original data?
5. Write an equation for your line.
6. Suppose a person who is absent from the class has a height of 160 cm. What does your line predict for his or her weight?
7. Repeat #6 for two other heights of your choice.
8. Find the mean of the heights, $\bar{x}$, and the mean of the weights, $\bar{y}$.
9. Plot the point with coordinates $(\bar{x}, \bar{y})$.
10. How close to your line is the point above?

Fig. 11.2

For example, students should recognize that a slope of 0.9 means that for every 10 centimeters taller a person is, he or she is about 9 kilograms heavier. Similarly, an intercept of -90 kg means that if this line were to make sense for a person with zero height, then his or her weight would be -90 kg. This leads to a discussion of appropriate and inappropriate extrapolations. Students can also compare their equations and note the variety. If they have estimated their lines well, they should also see that many are fairly close to the point representing the mean of each variable.

Finally, students can be presented with the question of determining who has found the "best" line. This can cause a bit of debate but eventually settles into questions of how close points fall to the line. Because slope and the structure of the linear equation relate changes in y to changes in x, we look at vertical distances of points from the line. After looking at absolute deviations from the fitted line, students are ready to consider a notion of squared deviations, the one generally used in formal regression. For support, it is helpful to make an analogy to the Pythagorean theorem. Here, too, sums of the squares create a simpler pattern than the distances themselves.

CALCULATOR ACTIVITY: THE LINE OF BEST FIT

Students are now ready to use a statistical graphing calculator to produce a line of best fit. They have encountered the intuitive underpinnings, understand the graphing considerations of designing an appropriate window, recognize the meaning and interpretation of such a line, have some insight into the theory, and are motivated to see how their lines compare with the calculator's output.

Using their height-weight data, students can use their calculators to generate a line of best fit and its parameters (see fig. 11.3 for sample output). Students need to be told that on the calculator, A represents the constant term and B represents the slope. Then, in groups, they can investigate other pairs of categories from the original measurement survey. Most data sets produce interpretable results. The pocket change data, however, create obvious anomalies. Usually the line produced is nearly horizontal. After some discussion, we see that the mean of the second variable is the best predictor when there really is no relationship between the variables; thus the horizontal line.

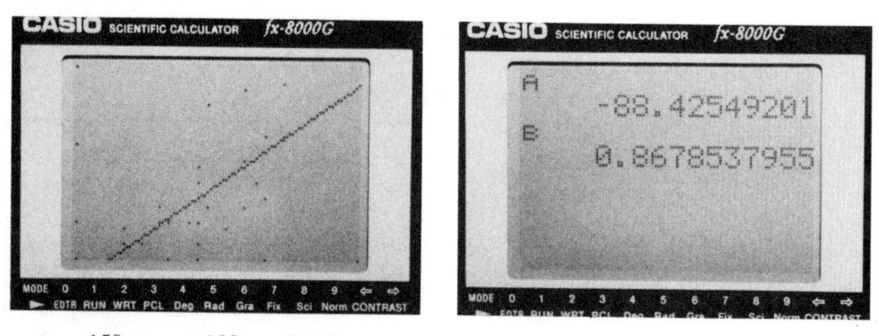

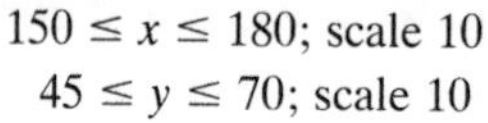
$150 \leq x \leq 180$; scale 10
$45 \leq y \leq 70$; scale 10

Fig. 11.3

Again, students can be challenged with more questions, as shown in figure 11.4. The last two questions connect data analysis to ideas of transformations. Students should realize that adding 2 to each x-value shifts the line 2 units to the right but does not affect the slope. Similarly, multiplying the weights (y) by a constant (0.9) shrinks the line vertically, multiplying both the slope and the y-intercept by 0.9.

Calculator Follow-Up Questions

1. Is the point $(\bar{x}, \bar{y})$ on all lines of best fit? Explain.
2. What happens when one of the variables is not related to the other variable?
3. Suppose all the students wore 2-cm lifts. Predict how this affects the line relating height and weight. Check your prediction by adding 2 to each height and producing a new line of best fit. Conjecture a generalization.
4. Suppose all the students took off 10 percent of their weight. Predict how this affects the line relating height and weight. Check your prediction by multiplying each weight by 0.9 and producing a new line of best fit. Conjecture a generalization.

Fig. 11.4

CORRELATION ACTIVITIES

By now students will have seen that although a set of bivariate data may have a single line of "best" fit, some lines fit their data sets better than others. They are ready to investigate correlation coefficients using small data sets like those in figure 11.5. Each team of students produces the scatter plot and correlation coefficient for a specified set. We compare the various displays and their respective r values, discuss the meaning of the correlation coefficients, and compare the relationship of these values to the corresponding graphs.

To emphasize the fact that correlation coefficients measure the linearity of a relationship, students can also explore data sets for points that fit perfectly on some known nonlinear curve—for example, $y = 12/x$, a hyperbola, or $y = -(x - 4)^2 + 11$, a parabola. They can use calculators to generate ordered pairs and scatter plots, then produce regression lines and correlation coefficients, and, finally, superimpose the graph of the original equation. The subsequent discussion should strengthen the idea that the calculator-produced correlation coefficient measures only how well data points fit one particular shape: a line that minimizes squared deviations.

At this point the calculators can be put aside and the students presented with situations having high correlations. The students are asked to discuss if one variable "causes" the other or if there is a more reasonable explanation. For example, children's height and reading scores correlate highly. Why? Both are functions of age! Students can concoct similar startling examples. Again, the message is reinforced that thinking plays a critical role in data analysis, not just number crunching.

Set A

x	y
1	2.0
2	2.5
3	3.0
4	3.5
5	4.0
6	4.5
7	5.0
8	5.5
9	6.0
10	6.5
11	7.0

Set B

x	y
2	3
3	2
4	3
4	5
5	7
6	6
7	3
7	6
9	6
9	8
11	9

Set C

x	y
1	5
3	3
4	5
4	9
6	3
6	8
7	5
7	11
9	7
11	8
11	11

Set D

x	y
1	4
1	7
1	10
3	2
4	9
6	11
7	3
8	8
9	5
11	3
11	11

Set E

x	y
1	9
3	3
3	6
3	8
5	8
6	3
7	9
9	5
10	2
10	7
11	5

Set F

x	y
1	10
2	5
2	8
3	6
4	4
4	7
6	5
6	8
7	2
8	3
11	2

Set G

x	y
1	6.00
2	5.75
3	5.50
4	5.25
5	5.00
6	4.75
7	4.50
8	4.25
9	4.00
10	3.75
11	3.50

Correlation coefficients are 1.00, 0.78, 0.49, 0.01, −0.35, −0.74, and −1.00, respectively.

Fig. 11.5. Data for correlation activity

EVALUATING MODELS

The next activity illustrates the way that "visual representation yields insights that often remain hidden from strictly analytic approaches" (Steen 1990, p. 8). Students find a line of best fit and the correlation coefficient for each of the four data sets in figure 11.6 (from Anscombe [1973]). Surprisingly, the results are the same for all four sets: regression lines are $y = 0.5x + 3$, and correlation coefficients are 0.82. Nevertheless, the graphs, as shown in figure 11.7, are markedly different. The first is a reasonable scatter plot of points modeled by a line. The second looks more like a parabola. The third is a perfect line except for one point; however, the "line of best fit" is not the line that includes the majority of the points. The fourth is a vertical set of points and one exceptional point, (19.0, 12.5). This last set shows dramatically how a single value can play a powerful role. As students tinker with the coordinates for this one point, they can see the effects on

Set 1

x	y
10.0	8.04
8.0	6.95
13.0	7.58
9.0	8.81
11.0	8.33
14.0	9.96
6.0	7.24
4.0	4.26
12.0	10.84
7.0	4.82
5.0	5.68

Set 2

x	y
10.0	9.14
8.0	8.14
13.0	8.74
9.0	8.77
11.0	9.26
14.0	8.10
6.0	6.13
4.0	3.10
12.0	9.13
7.0	7.26
5.0	4.74

Set 3

x	y
10.0	7.46
8.0	6.77
13.0	12.74
9.0	7.11
11.0	7.81
14.0	8.84
6.0	6.08
4.0	5.39
12.0	8.15
7.0	6.42
5.0	5.73

Set 4

x	y
8.0	6.58
8.0	5.76
8.0	7.71
8.0	8.84
8.0	8.47
8.0	7.04
8.0	5.25
19.0	12.50
8.0	5.56
8.0	7.91
8.0	6.89

Fig. 11.6. Four data sets

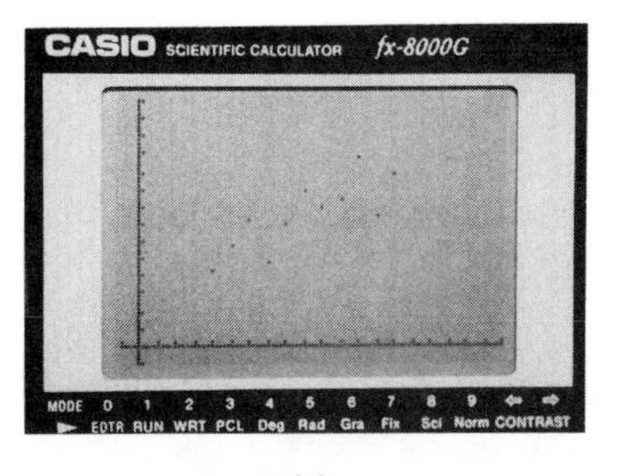

(a)

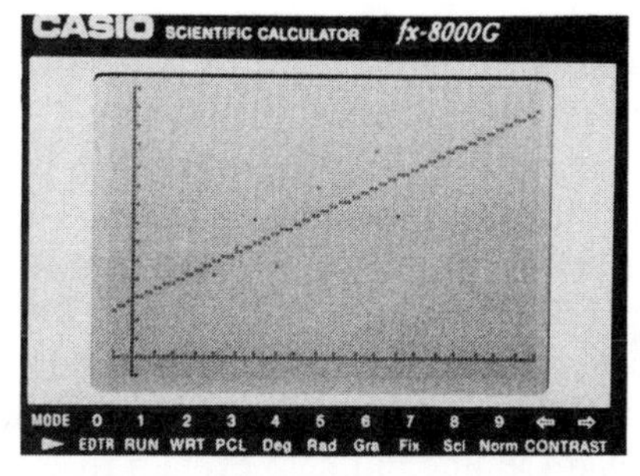

(b)

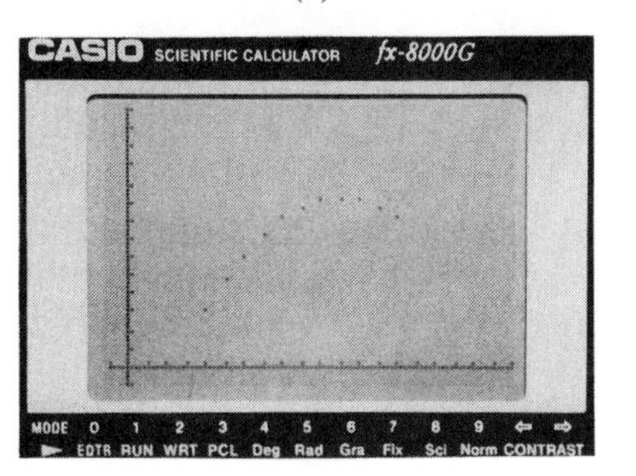

(c)

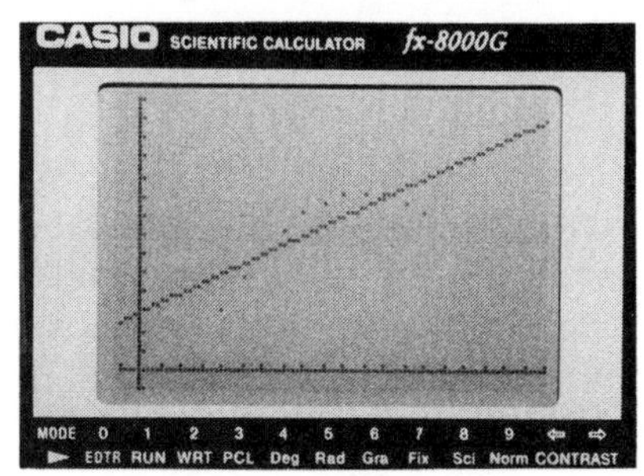

(d)

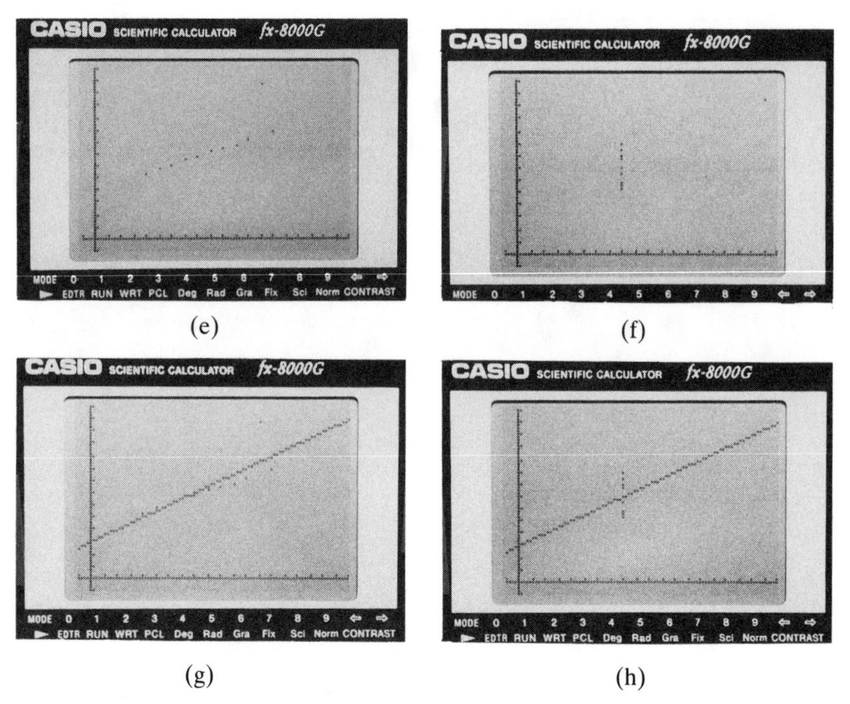

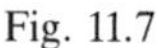
Fig. 11.7

the regression line produced by a single ordered pair. This activity impresses students with the value of looking at the *shape* of the data, not just the numerical parameters. It also underscores the usefulness of a graphing calculator as an investigative tool.

If technology is available with polynomial and other curve-fitting capabilities, then students can extend the idea of "best fit" beyond lines. For example, the data from set 2 in figure 11.6 can be entered and fit to a quadratic with a higher correlation coefficient than the linear model provides. Also, some of the measurement data can be re-explored. For example, a cubic can be used to estimate weight from height with a better fit than the line. To help explain this, students can be reminded that volumes of similar figures grow in proportion to the cubes of their lengths.

EXTENSIONS AND PROJECTS

Finally, students can do many independent or cooperative investigations. They can investigate bivariate data from almanacs, health journals, sports scoreboxes, product reports, and other sources. By using the calculator, students find that time spent on calculation is diminished and time spent on the analysis and interpretation is increased.

One interesting project involves looking at data sets with wide ranges versus truncated ranges. For example, students can compare correlation coefficients for height-weight data for students of one grade level against those for children and adults together. They can discover that truncated data sets give much weaker correlations than broad-based data sets do.

In a more advanced project, students can take several sets of data they have already studied and, for each, reverse the x- and y-values. Their experience with inverse functions suggests that the new line of best fit should be the inverse of the original one. However, they will discover that this is not so. The lack of symmetry is introduced because the least squares regression line measures distances from the vertical, not the shortest distance from a point to a line.

Another project compares hand-drawn or computer-generated median-median lines (Shulte and Swift 1986; Landwehr and Watkins 1986; North Carolina School of Science and Mathematics 1988) with lines of best fit. Students find that median-median lines are more resistant than least squares lines to outliers, particularly those near the extremes of the data.

Students who have studied exponential models can consider some population data, such as Canada's (as shown in fig. 11.8) (*Canadian Encyclopedia* 1988, p. 1720). Graphing the raw data suggests an exponential curve modeled by an equation of the form $y = Pr^x$, where P is a baseline population and r is the factor by which the population is multiplied in each time period measured by x. We transform the exponential curve to a line by taking the logarithm of the population values. This gives $\ln y = \ln P + x \ln r$, a linear equation in x. For simplicity we let 1851 be year 0, let x be the number of decades since 1851, and measure the population in millions. For a calculator with built-in exponential regression capability, data values can be entered directly. The first point, in this case, is (0, 2.4). For a calculator equipped

Year	Population
1851	2 436 297
1861	3 229 633
1871	3 689 257
1881	4 324 810
1891	4 833 239
1901	5 371 315
1911	7 206 643
1921	8 787 949
1931	10 376 786
1941	11 506 655
1951	14 009 429
1961	18 238 247
1971	21 568 311
1981	24 343 180

Fig. 11.8. Population of Canada

only with linear regression capability, the user takes the logarithm of each y value. The first point, in this case, is $(0, \ln 2.4)$. In the latter case, the parameters of the linear equation are $A = 0.9163$ and $B = 0.1757$, and the correlation is virtually 1. This corresponds to

$$\ln y = (0.9163) + (0.1757)x.$$

Exponentiating on both sides produces

$$y = (e^{0.9163})(e^{0.1757x}), \text{ or } y = 2.5\ (1.19)^x.$$

The expression above agrees with the output on a calculator producing exponential regression, as shown in figure 11.9. Again, students are asked, "What is the interpretation?" For 1851 ($x = 0$), the equation gives the

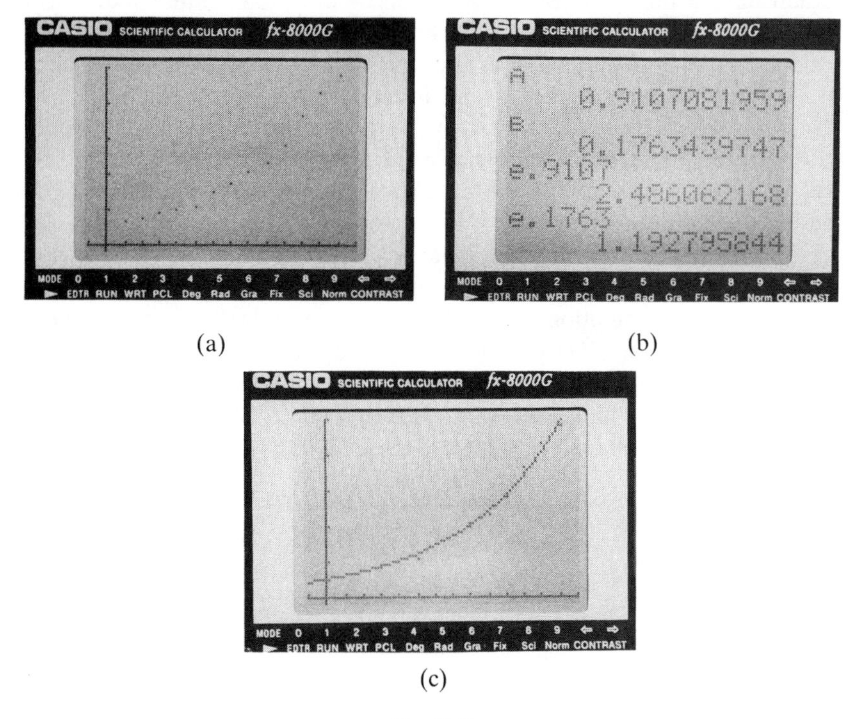

(a) (b)

(c)

Fig. 11.9

population as 2.5 million, very close to the actual number. For each successive decade, the population is multiplied by 1.19, a growth rate of about 19 percent per decade. Similar activities can be found in Rubenstein et al. (1992), Burrill et al. (1992), and North Carolina School of Science and Mathematics (1988, 1991).

SUMMARY

This article has illustrated a unit on line of best fit using statistical-graphing calculators. The activities blend measurement, statistics, algebra, by-hand graphing, and considerable discussion before, during, and after calculator activities. The calculator offers a number of educational advantages:

1. It makes topics formerly reserved for advanced study accessible earlier and to a broader range of students.
2. It provides a personal, portable tool that students can use in both independent and teacher-guided investigations.
3. It makes real data easy to handle.
4. It minimizes calculational asides, allowing more time for the development of concepts, translation of real situations into mathematics, and interpretation of results.
5. It opens opportunities for discovery learning and student projects.

With careful planning and some imagination, teachers can integrate calculators into instruction to achieve a variety of important improvements in mathematics education.

REFERENCES

Anscombe, Francis John. "Graphs in Statistical Analysis." *American Statistician* 27 (1973): 17–21.

Burrill, Gail, John C. Burrill, Pamela Coffield, Gretchen Davis, Jan de Lange, Diane Resnick, and Murray Siegel. *Data Analysis and Statistics across the Curriculum.* Curriculum and Evaluation Standards for School Mathematics Addenda Series, Grades 9–12. Reston, Va.: National Council of Teachers of Mathematics, 1992.

Canadian Encyclopedia. Edmonton, Alberta: Hurtig Publishers, 1988.

Landwehr, James M., and Anne Watkins. *Exploring Data.* Palo Alto, Calif.: Dale Seymour Publications, 1986.

National Council of Teachers of Mathematics. *Curriculum and Evaluation Standards for School Mathematics.* Reston, Va.: The Council, 1989.

North Carolina School of Science and Mathematics. *Data Analysis.* New Topics for Secondary School Mathematics. Reston, Va.: National Council of Teachers of Mathematics, 1988.

———. *Contemporary Precalculus through Applications.* Providence, R.I.: Janson Publications, 1991.

Rubenstein, Rheta N., James Schultz, Sharon L. Senk, Margaret Hackworth, John W. McConnell, Steven Viktora, Dora Aksoy, James Flanders, Barry Kissane, and Zalman Usiskin. *Functions, Statistics, and Trigonometry.* Glenview, Ill.: Scott, Foresman & Co., 1992.

Shulte, Albert P., and Jim Swift. "Data Fitting without Formulas." *Mathematics Teacher* 79 (1986): 264–71, 297.

Steen, Lynn A., ed. *On the Shoulders of Giants: New Approaches to Numeracy.* Washington, D.C.: National Academy Press, 1990.

12

Matrices + Modeling + Technology = Increased Mathematical Power for Students

Nancy Crisler
Gary Froelich

THE traditional approach to matrices at the secondary school level involves defining matrices and matrix operations in strictly mathematical terms. The operations of addition, subtraction, and scalar multiplication are easily mastered by students, but multiplication and matrix inversion are seen as difficult because they are time consuming and require elaborate calculations. Often the only application of matrices included in the curriculum is the solution of systems of equations. This approach results in students viewing matrices as another purely mathematical entity with little connection to the real world.

Calculators with matrix capabilities and computer software that performs matrix operations are commonly available. We have found that the availability of such devices in the classroom can result in greater student understanding of the logic of matrix operations and the usefulness of matrices in solving real-world problems. Meaningful applications no longer have to be avoided because of calculations that are too involved to perform by hand.

In this article we sketch an approach to matrices that emphasizes their role in modeling real-world phenomena. That such a modeling approach improves the problem-solving ability of students is well demonstrated by the success of secondary school students at the North Carolina School of Science and Mathematics in the Mathematical Competition in Modeling, a contest intended for college undergraduates.

MATRIX CALCULATORS

Features of calculators with matrix capabilities vary, but most allow addition, subtraction, scalar multiplication, multiplication, squaring, and inversion. Some perform matrix row operations. We have found that students often prefer to do addition, subtraction, scalar multiplication, and multiplication of small matrices by hand. They usually prefer the calculator for more involved multiplications, inversion, and row operations. We do not require students to demonstrate by hand proficiency at using algorithms for inverting matrices, multiplying matrices, or performing row operations.

ORGANIZING DATA WITH MATRICES

Initial experiences with matrices should entail the organization of data. Unwieldy collections of data are far less complicated when expressed in tabular form. Consider, for example, a retail clothing outlet that has opened stores in two malls. The outlet's November shipment of shirts from one of its suppliers was allocated in this way: 20 small to mall A and 18 small to mall B; 35 medium to mall A and 29 medium to mall B; 17 large to mall A and 14 large to mall B. The data are much easier to read if organized in a matrix:

$$\begin{array}{l} \\ \text{Small} \\ \text{Medium} \\ \text{Large} \end{array} \begin{array}{c} \text{Mall A} \quad \text{Mall B} \\ \begin{bmatrix} 20 & 18 \\ 35 & 29 \\ 17 & 14 \end{bmatrix} \end{array} \quad \text{November Allocation}$$

PROMOTING THE USE OF MATRIX OPERATIONS

Modeling settings allow students to see matrix operations as commonsense reflections of real-world activity. If the outlet receives a December shipment of shirts as shown in the matrix below, then determining the total number of small, medium, and large shirts allocated to the two stores in the months of November and December results in a natural definition of matrix addition.

$$\begin{array}{l} \\ \text{Small} \\ \text{Medium} \\ \text{Large} \end{array} \begin{array}{c} \text{Mall A} \quad \text{Mall B} \\ \begin{bmatrix} 11 & 17 \\ 28 & 27 \\ 14 & 12 \end{bmatrix} \end{array} \quad \text{December Allocation}$$

If the matrix shown below represents the total sales of shirts for the months of November and December, then determining the total number of small, medium, and large shirts that are left in stock at each store results in a natural definition of matrix subtraction.

$$\begin{array}{l} \\ \text{Small} \\ \text{Medium} \\ \text{Large} \end{array} \begin{array}{c} \text{Mall A} \quad \text{Mall B} \\ \begin{bmatrix} 22 & 24 \\ 44 & 39 \\ 20 & 17 \end{bmatrix} \end{array} \quad \begin{array}{c} \text{November and} \\ \text{December Sales} \end{array}$$

If the cost to the outlet of a small shirt is $9.20, the cost of a medium shirt is $9.55, and the cost of a large shirt is $9.90, then this information can be organized in a matrix in this way: [9.20 9.55 9.90]. Writing this matrix beside the November allocation matrix allows one to calculate the value of November's allocation for each store:

$$[9.20\ 9.55\ 9.90] \begin{bmatrix} 20 & 18 \\ 35 & 29 \\ 17 & 14 \end{bmatrix} = [686.55\ 581.15]$$

Prior to a formal description of matrix multiplication, students should be allowed to explore several problems like the previous one in order to properly understand such notions as the lack of commutativity of matrix multiplication and the importance of matching the number of columns of the first matrix with the number of rows of the second. We recommend introducing the matrix features of a calculator only after students have agreed on a proper definition of matrix multiplication. Prior to that, individual products and sums can be found by using the normal calculator arithmetic functions. For example, the key sequence 9.20 [×] 20 [+] 9.55 [×] 35 [+] 9.9 [×] [=] gives the product matrix entry 686.55, as required.

Additional information on representing data and promoting the use of matrix operations can be found in Froelich, Bartkovich, and Foerster (1991) and in North Carolina School of Science and Mathematics (1988).

A POWERFUL MODELING EXAMPLE

After students have experience with representing data in matrices and using matrix operations to solve problems, they should be exposed to problems that allow them to build matrix models and apply matrix operations.

Consider a species of wildlife whose numbers, birth rates, and survival rates are given by age group in the table below:

Age Group	0–5	6–11	12–17	18–23	24+
Current Population	18	25	0	14	0
Birth Rate	0	0.9	1.2	0.7	0
Survival Rate	0.5	0.9	0.9	0.7	0

How can the growth of the population from time period to time period be modeled? We allow students to explore this question by looking at the populations over several successive time periods. Students have access to calculators with matrix capabilities and are expected to use them to develop the matrix model.

The current population is represented in matrix T_0, and the birth and survival data are shown below in matrix L. Students often discover matrix

L when investigating the transition from one time period to the next. L is known as a Leslie matrix. Note that the total number of living members of the species, 57, is found by summing the members of matrix T_0.

$$T_0 = [18\ 25\ 0\ 14\ 0] \qquad L = \begin{bmatrix} 0 & 0.5 & 0 & 0 & 0 \\ 0.9 & 0 & 0.9 & 0 & 0 \\ 1.2 & 0 & 0 & 0.9 & 0 \\ 0.7 & 0 & 0 & 0 & 0.7 \\ 0 & 0 & 0 & 0 & 0 \end{bmatrix}$$

T_1, the population matrix at the end of the next time period, is found by performing the multiplication $T_0 \times L$. Of interest here is the overall growth rate of the species. One must first consider whether the total number of living members of the species is growing at a fairly constant rate. This can be done by considering the ratio of the number of living members between selected periods, such as between periods 28 and 29, and periods 29 and 30. Nearly equal ratios would be evidence of a fairly constant growth rate.

To know the total number of living members during time period 28 one must find matrix T_{28}, which is $T_{27} \times L$. Finding T_{28} by repeated multiplication would be time consuming, even with a matrix calculator. We have found that students confronted with this problem soon realize it can be solved by computing $T_0 \times L^{28}$. Although calculators allow only the squaring of matrices, there are reasonable ways around this limitation. On the TI-81, one calculates L^2, then ANS $\times$ L; and then repeats this calculation the desired number of times by repeatedly pressing the ENTER key.

In this example, the investigation of the numbers over several successive time periods shows a growth rate of about 9 percent. If the species under discussion is a game animal, this indicates that the harvesting of about 9 percent of the population by hunters would result in a steady population.

PROMOTING THE USE OF MATRIX ALGEBRA

In the next example we show how a modeling approach can be used to develop the notions of matrix inverse and matrix algebra. Consider a company composed of two divisions. One division manufactures electric-powered utility carts and the other manufactures batteries. Each division uses some of the product manufactured by the other. Each dollar's worth of product from the cart division requires 10 cents' worth of input from the cart division and 20 cents' worth of input from the battery division. Similarly, each dollar's worth of product from the battery division requires 4 cents' worth of input from the cart division and 6 cents' worth of input from the battery division. This information is represented in matrix A shown below:

$$A = \begin{array}{c c} & \begin{array}{cc} \text{Cart} & \text{Battery} \end{array} \\ \begin{array}{c} \text{Cart} \\ \text{Battery} \end{array} & \begin{bmatrix} 0.1 & 0.04 \\ 0.2 & 0.06 \end{bmatrix} \end{array}$$

We ask our students to determine the amount of product that is used within the company if the total output of the cart division is \$100 million and the total output of the battery division is \$90 million. If these figures are written in matrix O shown below, the result can be determined by calculating the product $A \times O$:

$$O = \begin{bmatrix} 100 \\ 90 \end{bmatrix} \qquad A \times O = \begin{bmatrix} 13.6 \\ 25.4 \end{bmatrix}$$

The product matrix, $A \times O$, indicates that \$13.6 million worth of product from the cart division was used within the company and \$25.4 million worth of product from the battery division was used within the company.

The difference between matrices O and $A \times O$ indicates the amount of product sold outside the company. Call this matrix D as shown below:

$$D = \begin{bmatrix} 86.4 \\ 64.6 \end{bmatrix}$$

If a market study indicates that the outside demand for the products of the cart division will be \$110 million and for the products of the battery division will be \$80 million, what amount of product should each division produce? In other words, if matrix D is known, what is matrix O? An algebraic analysis is shown below:

$$O - A \times O = D$$
$$(I - A) \times O = D$$
$$O = (I - A)^{-1} \times D$$

Here I represents the 2×2 identity matrix, $\begin{bmatrix} 1 & 0 \\ 0 & 1 \end{bmatrix}$, and $(I - A)^{-1}$ is the inverse of the matrix found by subtracting matrix A from the identity matrix. Students better understand the logic of matrix algebra if it is compared to that of the algebra with which they are familiar. Compare, for example, the solution of this matrix equation to solving the algebraic equation $x - 3x = 7$ by factoring the terms on the left rather than combining them.

Calculators with matrix capabilities can evaluate the expression on the right once matrices A, D, and I are entered. Some calculators limit the number of matrix calculations in a single expression. In that instance, $(I - A)^{-1}$ would be evaluated first, and then the answer multiplied by D.

A BRIDGE FROM DATA TO FUNCTIONS

The traditional application of matrices to the solution of systems of equations is made even more appropriate by matrix calculators. In the next example we demonstrate a connection between algebra, matrices, and data analysis.

Figure 12.1 is a photograph of a golf ball thrown upward from a point near the lower left edge of the photograph. It was taken against a dark background in a darkened room with a one-second exposure setting on a

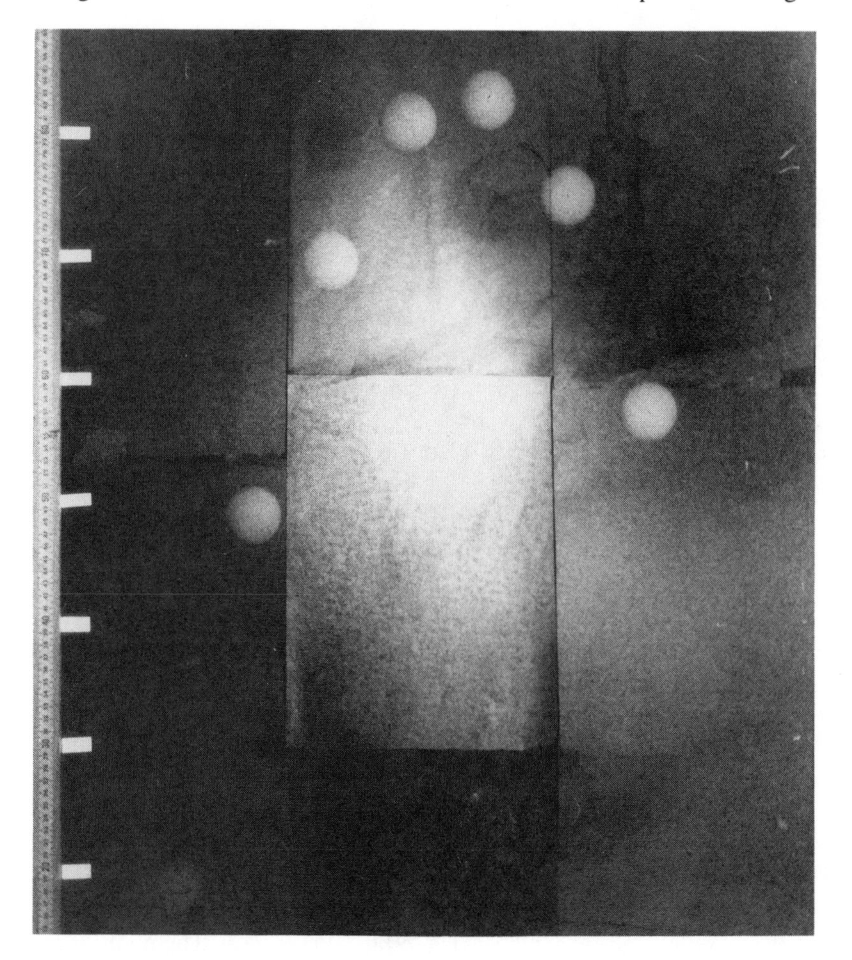

Fig. 12.1

35-mm camera. Illumination was provided by a strobe light borrowed from a school physics labs. A meterstick was taped to the wall to allow readings of heights.

The first column in table 12.1 contains the number of strobe flashes (time units), and the second column contains the height in centimeters of the center of the ball from the bottom of the stick. An 8 × 10 enlargement of the photograph was used to improve the accuracy of measurements.

Table 12.1

Flash Number	Height	First Difference	Second Difference
1	48.7		
		20.8	
2	69.5		−9.6
		11.2	
3	80.7		−9.5
		1.7	
4	82.4		−9.3
		−7.6	
5	74.8		−9.7
		−17.3	
6	57.5		

The third column of the table shows the differences in successive heights. The nearly constant second differences shown in the fourth column suggest that a quadratic function, $f(t) = at^2 + bt + c$, would fit the data well.

The quadratic model and the first, fourth, and sixth data points yield this system of equations:

$$\begin{aligned} 48.7 &= a + b + c, \\ 82.4 &= 16a + 4b + c, \\ 57.5 &= 36a + 6b + c. \end{aligned}$$

The system can be solved with a matrix calculator that performs inversion and multiplication by multiplying the inverse of the matrix of coefficients by the column matrix of constants:

$$\begin{bmatrix} 1 & 1 & 1 \\ 16 & 4 & 1 \\ 36 & 6 & 1 \end{bmatrix}^{-1} \times \begin{bmatrix} 48.7 \\ 82.4 \\ 57.5 \end{bmatrix} = \begin{bmatrix} -4.74 \\ 34.92 \\ 18.52 \end{bmatrix}$$

Thus, the function is $d(t) = -4.74t^2 + 34.92t + 18.52$. It can be demonstrated that the function fits the data well by checking it against other values in the table: $d(2) = 69.4$, $d(3) = 80.6$, $d(5) = 74.6$. The photograph contains a position of the ball that did not catch enough light from the strobe to be easily seen. Possible locations can be determined by evaluating the model at $t = 0$ and $t = 7$.

SUMMARY

In summary, we have considered a number of applications that make matrices meaningful to students. Other topics that reflect a modeling approach include linear programming, Markov chains, game theory, and graph theory. See the report of the NCTM Task Force on Discrete Mathematics (1990) for additional information. A novel application of matrices that appeals to students exists in cryptography, the science of coding and decoding information (see Malkevitch and Froelich [1991]). The computations required for these applications make their study prohibitive to students who must perform the computations with pencil and paper alone. Calculators with matrix functions, however, make these exciting real-world problems accessible to secondary school students.

REFERENCES

Froelich, Gary, Kevin Bartkovich, and Paul Foerster. *Connecting Mathematics*. Curriculum and Evaluation Standards for School Mathematics Addenda Series, Grades 9–12. Reston, Va.: National Council of Teachers of Mathematics, 1991.

National Council of Teachers of Mathematics, Task Force on Discrete Mathematics. *Discrete Mathematics and the Secondary Mathematics Curriculum*. Reston, Va.: The Council, 1990.

Malkevitch, Joseph, and Gary Froelich. *Codes Galore*. Arlington, Mass.: Consortium for Mathematics and Its Applications, 1991.

North Carolina School of Science and Mathematics. *Matrices: Materials and Software*. New Topics for Secondary School Mathematics. Reston, Va.: National Council of Teachers of Mathematics, 1988.

13

Using Graphing Calculators to Investigate a Population-Growth Model

Rosalie Dance
Zachary Jeffers
Joanne Nelson
Joan Reinthaler

THROUGH the use of graphing calculators, the mathematics associated with many real-world problems is accessible to high school students. This technology permits the investigation of dynamical systems through both symbolic and graphic representations of difference equations and recursion without the drudgery that has been a barrier in the past. Such a study can develop an intuitive basis of the limit concept and provide a background for further study in a calculus course. It is also now feasible to recommend "investigating" a problem rather than focus on "solving" a problem. It is in such investigations that much of the creative work of modeling is accomplished.

BUILDING MODELS, BUILDING CONCEPTS

The article in figure 13.1 from the *Washington Post* about controlling the deer population at three reservoirs in Baltimore, Maryland, prompted considerable discussion in a high school mathematics class and, subsequently, the development of mathematical models of population growth and control. The analysis of such models is uniquely suited to the capabilities of graphing and programmable calculators.

Students are easily convinced that the size of a population in any year depends on the size of the population in the previous year, and it is not

---

Much of this paper was written during the 1989 NSF-funded workshop, "Modeling with Discrete Mathematics," which was held at Georgetown University. We are grateful to project directors James Sandefur and Monica Neagoy for their help and encouragement.

Controlling Md. Deer

■ Maryland wildlife officials have asked the City of Baltimore to consider opening three reservoirs to firearms hunting to resolve a situation brought on by an increasing deer population.

Deer at the Loch Raven Reservoir are leaving to look for food, and neighbors are complaining that they are eating shrubbery and gardens.

The deer are also making it virtually impossible for the city to reforest the watershed areas, and are eating underbrush that serves to filter the water that flows into reservoirs.

Mayor Kurt L. Schmoke said a task force is working on the problem. Some options include feeding the deer or having an expanded hunting season.

"The only feasible way we have of controlling the population is by killing the animals," said John Sandt, forest wildlife supervisor for the state Department of Natural Resources. "And we can either do it with sharpshooters or through the hunting public."

"We have gotten a few calls from people," said Walter Koterwas, chief of the city's water facilities. "Some say, 'Eliminate the pest, period,' and others say, 'We have to save Bambi.' "

Fig. 13.1. 1990, *The Washington Post.* Reprinted with permission.

necessary, in constructing a descriptive model, to consider all the variables (e.g., food supply, predators) that influence this dependence. In its simplest form, it can be argued that the size of the population in year $n + 1$, which we shall denote $A(n + 1)$, is some constant multiple, $(1 + r)$, of $A(n)$, where r is the growth ratio and $A(n)$ is the size of the population in year n. This gives rise to the recursive model

$$A(n + 1) = (1 + r)A(n), \text{ or } A(n + 1) = r(A)n + A(n),$$

whose solution is

$$A(k) = A(0)\,(1 + r)^k.$$

Using a calculator, students can explore this model. With a graphing calculator such as the Casio fx – 7000 or TI-81, one might enter a value for $A(0)$ and press [EXE]. Then, taking a value for r, type $(1 + r)$ [ANS] and [EXE]. This gives $A(1)$. To continue, just press the [EXE] key repeatedly. If students study the table they can produce in this way, they will probably conclude that this model is not entirely satisfactory (see table 13.1).

Table 13.1

$A(n + 1) = (1 + r)A(n)$ where $A(0) = 150$ and $r = 0.5$

n	$A(n)$
0	150
1	225
2	337.5
3	506.25
4	759.38
5	1139.06
6	1708.59
7	2562.89
8	3844.34
9	5766.50
10	8649.76

Populations do not and cannot grow exponentially forever. The environment itself has some limited capacity that a population cannot exceed.

Therefore the growth ratio, r, cannot be invariant. It might be supposed that the growth ratio itself depends on the carrying capacity of the environment (which we shall call L) and that as the population nears L, the growth ratio decreases. This requirement can be modeled by introducing the factor $(1 - A(n)/L)$ into the equation. Notice that as $A(n) \to L$, $(1 - A(n)/L) \to 0$, and when $A(n)$ is small, $(1 - A(n)/L)$ is close to 1. The new model looks like this:

$$A(n + 1) = r(1 - A(n)/L)A(n) + A(n)$$

Investigation with a calculator as described above shows students that the new model behaves as we hoped. For small populations, growth is more rapid. As the population nears L, growth slows down and the population curve is bounded above asymptotically by the value L for reasonably small r values.

A particular population can now be modeled by assigning values to the constants. Consider an environment, with an initial deer population of $A(0) = 150$ and a growth ratio $r = 0.5$, that can support a population of $L = 2000$. The equation becomes

$$A(n + 1) = 0.5(1 - A(n)/2000)A(n) + A(n).$$

The deer population model can be interpreted as the iteration of a function, the composition of the function with itself. The output of the function becomes the input of the function at the next step. It has been observed that the number of deer present after a given unit of time is a function of the number present at the beginning of that unit of time: $A(n + 1) = f(A(n))$. Iterating the function permits the student to predict the population after n time periods have passed. Later it might be useful to build a model that includes other control factors. By iterating the new function, the student can test a new hypothesis.

Under what circumstances does the deer population reach equilibrium (i.e., settle down and just maintain itself)? When the "input" is equal to the "output." This can be visualized as the point where the function $A(n + 1) = f(A(n))$ intersects the function $A(n + 1) = A(n)$.

This may be easier to understand in conventional algebraic notation. A function $y = f(x)$ has a fixed point under composition with itself where its graph intersects the graph of $y = x$.

The function $f(x) = \cos(x)$ provides a pretty example. Have students put their calculators in radian mode and then enter any number and take its cosine. Then take the cosine of the output and repeat this many times. On most standard scientific calculators and on the Hewlett-Packard 28S, this is accomplished by entering a first number and then pressing [COS] repeatedly. On the Casio fx – 7000G and on the TI-81, it is best to enter a number first, press [EXE], then press [COS] [ANS], and finally press [EXE] repeatedly.

This process evaluates the cos(cos(cos(. . . (*x*)))). Most people are amazed to discover that the calculator eventually settles down to producing the same number over and over, approximately 0.739. When the iteration process achieves this equilibrium, we have found the *fixed point* of the function $f(x) = \cos(x)$.

A graphical image of this process is called a *cobweb*. It is illustrated in figure 13.2 and can be produced using the following steps:

Graph $f(x) = \cos(x)$ using the same scale on both axes. On the same coordinate system, graph $y = x$.

1. Choose initial input, x_0
2. Get $y = \cos(x_0)$
3. Locate a point whose *first* coordinate is the *y*-value you just got. (On the graph, move horizontally to $y = x$.)
4. Find the function value (*y*-value) for this new *x* by moving vertically.
5. Repeat step 3.
6. Repeat step 4.
7. Repeat step 3.
8. Repeat step 4 and continue.

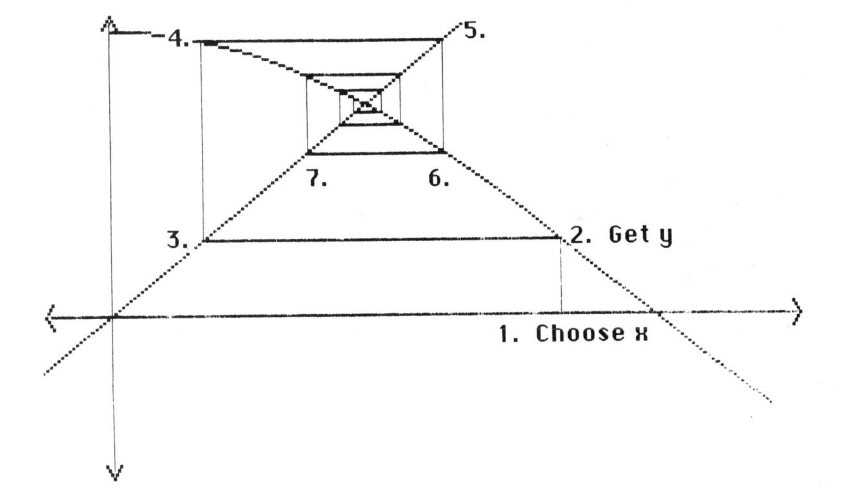

Fig. 13.2

It is important to emphasize to students that this is not a different process from their earlier calculator search for a fixed point but a graphical image of that search. The relevance of the line $y = x$ in graphing recursion reinforces its significance for inverses of functions under the operation of composition.

For a second example, students can easily make a cobweb of $f(x) = (x + 7)/3$ (fig. 13.3). Geometrically, this is quite different from the cobweb for $f(x) = \cos(x)$, but the process is the same.

Remember, from the point $(x_0, 0)$ move first vertically to the line $y = f(x)$, then horizontally to the line $y = x$, then vertically to $y = f(x)$, and so on. In this way, output is converted to input, and a new output is then computed.

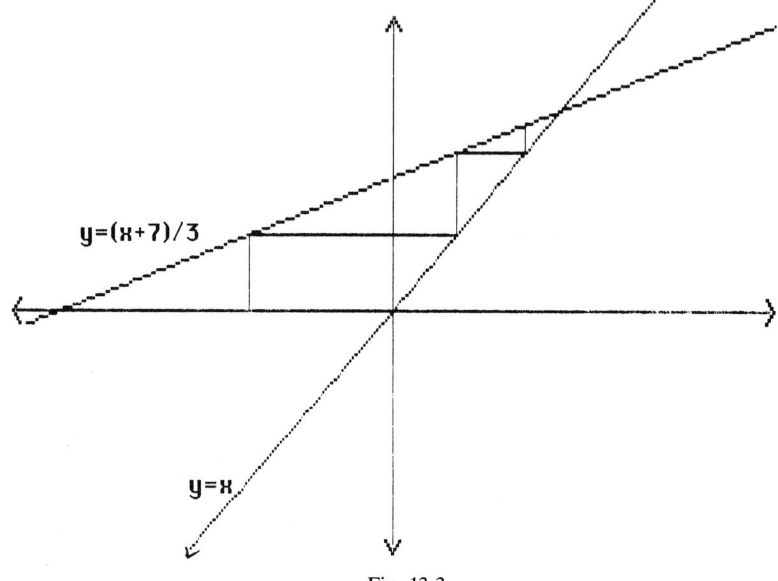

Fig. 13.3

Cobwebs of linear functions with varying slopes will lead students to note for themselves the categorization of fixed points into two different kinds: attracting fixed points and repelling fixed points. Sufficient exploration will probably lead students to conjecture the following: (1) when the slope, m, of the graph is negative, the cobweb "spirals," and when m is positive, the cobweb "stairsteps"; (2) when $|m| < 1$, the fixed point is attracting, but when $|m| > 1$, the fixed point is repelling.

A small amount of practice in doing graphical analysis by hand is valuable. But both in reducing tedium and in increasing accuracy, the graphing calculator has significant advantage. Most graphing calculators permit plotting points and producing line segments. This is all that is needed to draw a cobweb.

On a Casio graphing calculator the following simple program can be stored and then called after graphs are plotted:

```
Line:Plot X,Y
```

After graphing both the function to be analyzed and the identity function, call this program. Using the cursor keys, move the cursor to the $A(0)$ point on the x-axis. Press EXE. Move the cursor vertically to the function to be analyzed. Press EXE. Move the cursor horizontally to the line $y = x$. Press EXE, and continue.

On the TI-81, after graphing the function to be analyzed and the identity function, press <2nd> DRAW 2 to begin the cobweb. Use the arrow keys to move the cursor to the $A(0)$ point on the horizontal axis. Press ENTER. Move the cursor vertically to the function to be analyzed. Press ENTER twice (first to end the first segment, then to begin the next one). Move the cursor horizontally to meet $y = x$. Press ENTER twice, and continue.

At this point, one might wonder what deer populations have to do with the composition of functions and with cobwebs. The recursive model of population growth that has been developed is

$$A(n + 1) = 0.5(1 - A(n)/2000)A(n) + A(n).$$

Suppose a pair of axes is drawn. The horizontal axis, the x-axis, represents "input"; the vertical axis represents "output." To plot a graph of the population function, then, let $y = A(n + 1)$ and $x = A(n)$. Thus, graph $y = f(x) = 0.5(1 - x/2000)x + x$ on the x,y plane. Simplifying this yields the function

$$y = f(x) = -0.00025x^2 + 1.5x.$$

The graph of this function is a parabola; with some simple factoring, it is found that its zeros are $x = 0$ and 6000 and its vertex is at the point (3000, 2250).

If this function is graphed on the same axes as the graph of $y = x$ (fig. 13.4), the points of intersection represent the fixed points, points where $A(n + 1) = A(n)$. The coordinates of these points of intersection can be determined by using the calculator's trace function. (The x- and y-values given by the calculator are approximations.)

The behavior of the population can now be investigated using cobwebs. The beginning population, $A(0)$, is 150 deer. This should be the starting point (see fig. 13.5). A vertical line through $x = 150$ intersects the parabola at a point whose second coordinate, $y = f(150)$, represents the population at the end of the first year. $A(0)$ is the input, and $A(1)$ is the output. Now a horizontal line through $[A(0), A(1)]$ intersects the line $y = x$ at the point $[A(1), A(1)]$, and if a vertical line is now drawn through $[A(1), A(1)]$ to the point $[A(1), A(2)]$ on the parabola, $A(1)$ becomes the input and $A(2)$ is the new output. The line $y = x$ is the mechanism by which the old output becomes the new input. The cobweb tracks the path of this recursion. It

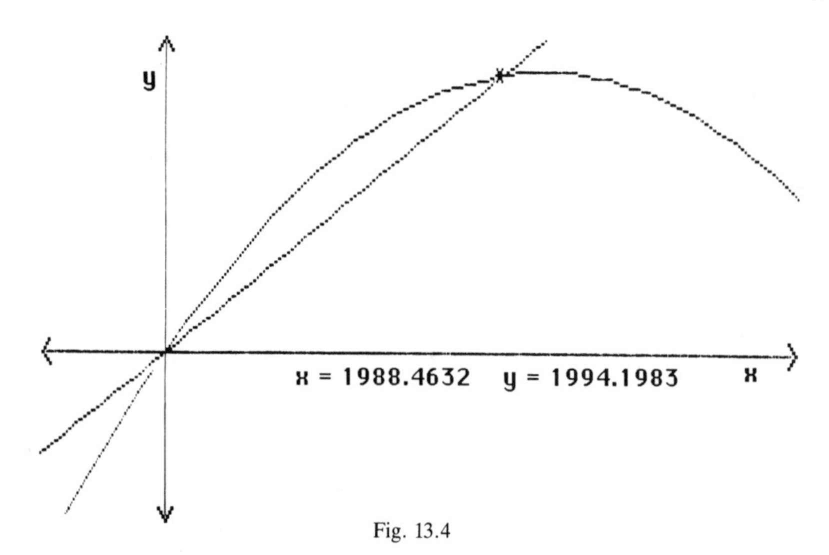

Fig. 13.4

should be remembered that the value of each $A(n)$ represents the population at year n.

On the graphing calculator this is accomplished by moving the cursor from $A(0)$ on the horizontal axis vertically to the parabola and drawing the segment, then moving the cursor horizontally to the $y = x$ line and drawing a segment, and so on.

It can be seen that as time goes on, the population approaches the carrying capacity of 2000, the first coordinate of the upper fixed point.

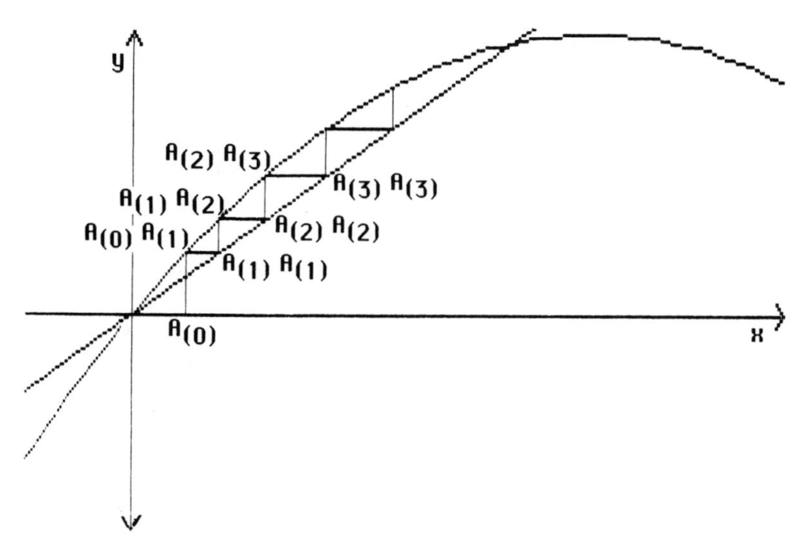

Fig. 13.5

HARVESTING

The controversy reflected in the news article (fig. 13.1), however, centers on ways to *control* population growth above and beyond the limitations imposed by the environment itself. Euphemistically known as "harvesting," this can take many forms. One simple system, and the one considered here, is to kill off a fixed number (h) of animals each year. This requires that a term, $-h$, be added to the previous model, giving

$$A(n+1) = r(1 - A(n)/L)\,A(n) + A(n) - h,$$

where h is some constant. This harvesting model leads to a new recursion relationship that can be investigated using cobwebs.

Modifying the previous model, where $r = 0.5$ and $L = 2000$, to include a fixed harvest of h animals each year yields the new relation

$$A(n+1) = 0.5(1 - A(n)/2000)\,A(n) + A(n) - h.$$

If $h = 68$, then $A(n+1) = 0.5(1 - A(n)/2000)A(n) + A(n) - 68$.

This population will stabilize where $A(n+1) = A(n)$. If $A(n)$ is replaced with x, the equation for equilibrium becomes

$$x = 0.5(1 - x/2000)x + x - 68,$$

which simplifies to

$$0 = -0.00025x^2 + 0.5x - 68.$$

The roots of this equation are approximately 147 and 1853.

At this point, it would be useful to investigate the attracting or repelling nature of the fixed points. Using a cobweb program for a graphing calculator (fig. 13.6), experiment with different initial values of $A(0)$. The cobweb graph for $A(0) = 150$ is shown in figure 13.7. What do you notice about the cobwebs when the value of $A(0)$ is between 147 and 1853? When the value of $A(0)$ is greater than 1853? When $A(0)$ is less than 147? The fixed point with $A(n) = 1853$ is called an *attractor,* and the point with $A(n) = 147$ is called a *repeller.*

By observation one can discover that a fixed point is an attractor if the slope of the graph at that point has an absolute value less than 1 and a repeller if the slope's absolute value is greater than 1. In some courses it may be possible to use calculus to do a more rigorous analysis of the nature of fixed points.

Now, changing the value of h translates the graph of the recursive relation up or down and changes both points of the graph's intersection with the line $y = x$ (fig. 13.8). As more deer are harvested, the parabola is translated down and the points of intersection move closer together. As fewer are

Cobweb Program for Casio Calculators

```
Cls: "X1" ? → A: "X2" ? → B: Range A,B,1,A,B,1:
"A(0)" ? → C: "N" ? → N: "A(N+1)=RA²(N)+SA(N)+T,
R=" ? → R: "S=" ? → S: "T=" ? → T:
Graph X: Graph RX²+SX+T: Plot C,0:
Lbl 1: RC²+SC+T→D: Plot C,D: Line:
Plot D,D: Line: D → C: Dsz N: Goto 1:
```

Cobweb Program for TI-81 Calculators

```
:ClrDraw :All-Off :Disp "X₁" :Input A :Disp
"X₂":Input B :A→Xmin :A→Ymin :B→Xmax
:B→Ymax:Disp "A(0)" :Input C :Disp "N"
:Input N :Disp "A(n+1)=RA²(N)+SA(N)+T, R="
:Input R : Disp "S=" :Input S :Disp "T"
:Input T :DrawF X :DrawF RX²+SX+T :Lbl 1
:RC²+SC+T→D :Line(C,C,C,D) :Line(C,D,D,D)
:Pause :D→C :DS<(N,0) :Goto 1
```

In both programs, X_1 is the lower boundary of the domain and range value; X_2 is the upper boundary of the domain and range value; $A(0)$ is the initial value; N is the number of iterations; and R, S, and T are the coefficients of the function $RX^2 + SX + T$.

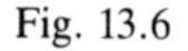

Fig. 13.6

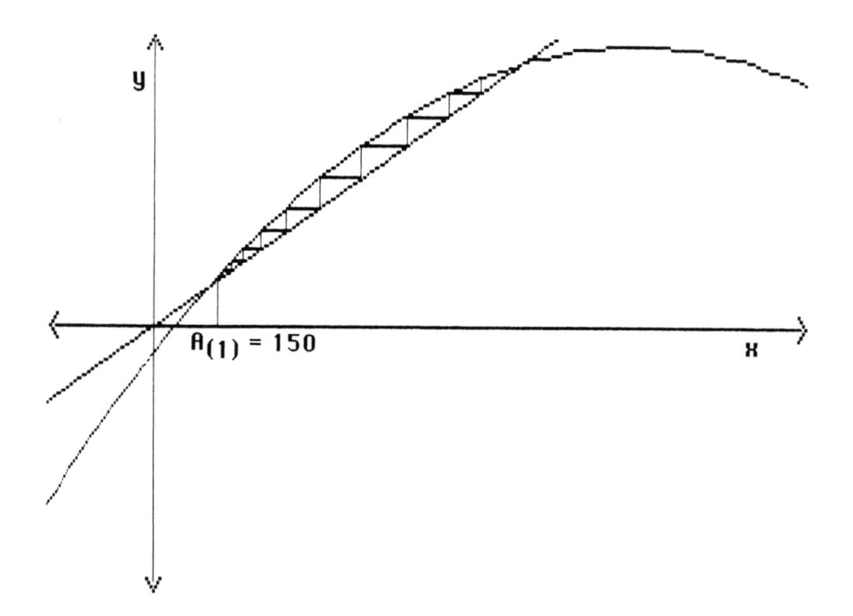

Fig. 13.7

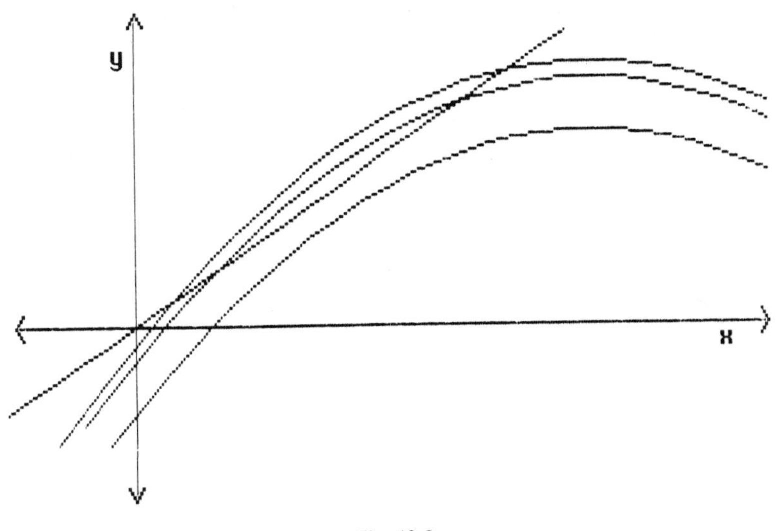

Fig. 13.8

harvested, the points move farther apart and approach their limiting values of 0 and 2000.

As long as $A(0)$ is a value between the two equilibrium points, the population will tend to the upper point. If h and $A(0)$ are such that $A(0)$ is less than the lower equilibrium value, however, the population will die out. If h is such that there are no equilibrium points (the two graphs do not intersect), then the population always dies out. If h is such that there is exactly one equilibrium point (the graphs are tangent), then the population survives only when $A(0)$ is the first coordinate of the point of tangency. In this case the value of h is the maximum possible number of deer that can be harvested without having the population die out.

Other harvesting systems can be investigated using a graphing calculator. In particular, a system in which a population is "managed" by harvesting a fixed *proportion* of the population each year has the recursive model

$$A(n + 1) = 0.5(1 - A(n)/2000)A(n) + A(n) - rA(n),$$

where r is some fixed proportion. It might be useful to investigate the ways that this model differs from the fixed-harvest model.

SUMMARY

The application of the equilibrium concept to the problem of deer population growth provides an opportunity for students to experience building an equation that models a real situation. Real-world applications of iteration abound: compound interest and amortization, radioactive decay, Newton's method for finding the roots of an equation, and fractals. Observing the relationship between a first-order affine dynamical system and the exponential function extends students' understanding of exponential function; observing the nonlinear dynamical system with no general solution challenges them to move beyond cut-and-dried formulas. Cobwebs provide an opportunity to develop visually an intuitive understanding of limits.

The use of the calculator in producing data to model a problem permits the investigation of mathematical topics that would otherwise be inaccessible to high school students. Aside from data production, the use of the graphing calculator aids both in developing concepts and in solving problems.

BIBLIOGRAPHY

McDonald, J. J. "Deer Harvesting." In *Case Studies in Mathematical Modeling*, edited by D. J. G. James and J. J. McDonald, pp. 27–54. New York: John Wiley & Sons, 1981.

Piel, E. J., and J. G. Truxal. *Man and His Technology.* New York: McGraw-Hill Book Co., 1973.

———. *The Man-made World.* New York: McGraw-Hill Book Co., 1971.

Sandefur, James T. *Discrete Dynamical Systems: Theory and Applications.* London: Oxford University Press, 1990.

14

Graphical Insight into Elementary Functions

Judith H. Hector

EASY-TO-USE computer programs that graph functions have been available for several years. More recently, the price of graphing calculators has dropped to the point where students can afford to purchase their own. Such graphing devices can have an important effect on how elementary functions are taught. This article focuses on the benefits of using calculators to graph and also gives some specific suggestions on how to use these devices in the teaching of elementary functions.

The most obvious benefit of graphing calculators is the ease with which they produce complete pictures for a variety of functions. Without a graphing calculator, students must construct a table of $(x,f(x))$ values and plot those points to produce a hand-drawn graph. However, calculating a few points can be laborious and may not lead to an accurate picture of a function. Graphing calculators allow students to graph a function quickly, to isolate a section of the graph, to zoom in for greater detail, to zoom out to view the function as x increases or decreases, and to compare the graph of one function with the graph of another function.

FAMILIARITY WITH GRAPHICAL REPRESENTATIONS—CLASSIFICATION

With little training, students using graphing calculators can graph families of curves and discover the connections between algebraic and graphical representations of functions. In first-year algebra, an appropriate family to begin the exploration of linear functions might include $y = x$, $y = x + 5$, and $y = x - 2$. If students graph these three lines on the same screen and describe what they see, the concept of the y-intercept emerges naturally.

A second family of lines allows students to investigate slope. For example, students readily note that $y = x$, $y = 2x$, and $y = 3x$ appear as lines with increasingly steep slopes. If students then graph $y = -3x$, they see a line with a left-to-right downward slope. This kind of exercise makes students active participants in graphing. Students using calculators enthusiastically observe and generalize from the graphs they produce.

Initially, the teacher may specify appropriate minimum and maximum x- and y-values for the graphs, or students may use a default viewing window. As students gain more experience, they need to work with functions that do not fit a default window. When functions arise from application problems, domains may have restrictions such as $x \geq 0$. Small values of x, say between 0 and 5, may produce functional values in the thousands. Students learn about functional behavior as they search for an appropriate domain and range to give a full view of the function.

When students work with different types of functions and produce graphs of each type, they become familiar with the relation between function rules and graphs and learn to classify them as linear, quadratic, higher-degree polynomial, rational, exponential, logarithmic, or trigonometric. Students gain greater familiarity with quadratic functions from examining examples like $y = (x - 2)^2$, $y = (x - 0)^2$, and $y = (x + 5)^2$. Graphs of the three parabolas on one viewing screen help students see the common shape and shifts left and right. A next step is to ask what happens when the coefficient of x^2 is varied, as in $y = 0.5x^2$, $y = x^2$, and $y = 3x^2$. Similar explorations are fruitful with $y = a^{bx}$ where a and b are varied. The graphs of $y = 3^{bx}$, where $b = 1, 2$, and 3, are shown in figure 14.1.

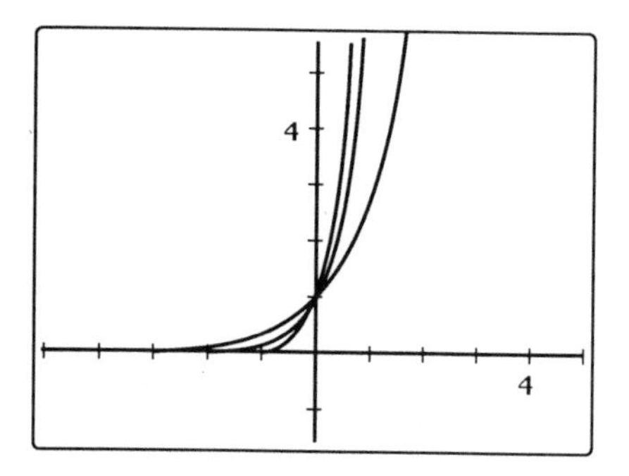

Fig. 14.1

Individually or in small groups, students can be asked to graph, observe, discuss, and write down observations. Observations can be shared with the class and conclusions refined. Certainly textbooks and teachers attempt similar graphical presentations. For example, teachers may sketch $y = \sin x$ and contrast the graph with $y = \cos x$. Textbooks may show graphs of $y = \sin x$ and $y = \sin 2x$. However, an investigation in which students work with the general form $y = a \sin (bx + c) + d$ and choose different values of a, b, c, and d to see how the graphs vary has strong impact because it asks students to discover the relationships between graphical and symbolic representations.

The aim is to have students develop a rich base of experiences with the rules and graphs of elementary functions and to recognize which rules go with which graphs. These experiences form a firm foundation for later work with limits, continuity, derivatives, area under a curve, and other calculus concepts that can be introduced graphically.

GRAPHING IN CONJUNCTION WITH ALGEBRAIC TECHNIQUES

Graphical approaches to functions can also clarify algebraic methods. For example, consider the teaching of quadratic inequalities. The algebraic solution of $x^2 - 5x - 6 > 0$ may consist of factoring to the form $(x - 6)(x + 1) > 0$. Then it is noted that the inequality is true when the two factors are both positive or when they are both negative. One method of solution involves testing values to the left of -1, between -1 and 6, and to the right of 6. Alternatively, another method examines unions and intersections with the inequalities

$$x - 6 > 0 \text{ and } x + 1 > 0 \quad \text{or} \quad x - 6 < 0 \text{ and } x + 1 < 0.$$

Unfortunately, some students see only the possibility on the left, both factors positive. They may reason that since the original inequality used "$>$," they should restrict their attention to this inequality throughout the rest of the algebraic work.

With a graphing calculator, it is possible to begin with the graph of the function $y = x^2 - 5x - 6$, shown in figure 14.2. The question to be answered graphically is as follows: For what x-values is y above the x-axis? Students can see that the form of the answer will be x-values to the left of one zero of the function and x-values to the right of the other zero of the function. A focus on symbolic methods usually makes no such connection between quadratic inequalities and graphs of quadratic functions.

With a graphing calculator, it is also feasible to introduce cubic or higher-degree inequalities graphically. An example of a cubic inequality, $(x - 3)(2x - 1)(x + 4) < 0$, is shown in figure 14.3. Students can estimate the solution, $x < -4$ and $0.5 < x < 3$, by reading the scale markings. Without a graphing calculator, the algebraic solution of cubic or higher-degree inequalities usually comes later, if at all, in the instructional sequence. However, locating on a graph where a function is above or below the x-axis is only a little more difficult for higher-degree functions than for linear or quadratic ones.

The main purpose here of the graphical approach is not to replace an algebraic approach but to clarify it. Without the ability to view a graph

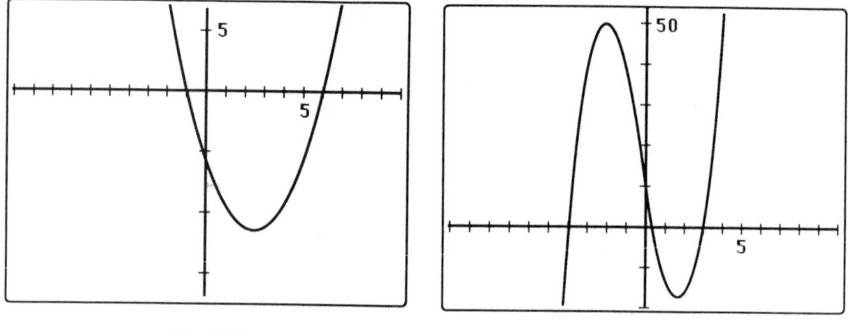

Fig. 14.2

Fig. 14.3

quickly, the algebraic process of testing values in an inequality can be confusing, and students have no easy way of judging whether they have made a mistake.

FINDING ROOTS

Root finding is required by an important collection of problems in mathematics, and there are many important techniques to accomplish that task. High school students are well rehearsed in techniques for solving linear and quadratic equations, but they do much less with other classes of equations. Many are baffled by the solution of trigonometric equations. First, it is not clear that students distinguish between verifying a trigonometric identity and solving a trigonometric equation. Second, students are usually amazed when the answer key lists multiple roots for an equation such as $3 \tan^4 x = 1 + \sec^2 x$ when it is solved for x in the interval $[0, 2\pi]$.

To help students distinguish between trigonometric identities and equations, a graphing calculator is valuable. As a classroom activity, students may be grouped in pairs. Then one student graphs the right side of an identity as a function, and the other student graphs the left side. As the pairs compare their graphs, students should see that they are identical. Even in the case of a fundamental identity such as $\sin^2 x + \cos^2 x = 1$, students may not have recognized that the left side of the identity is a constant function. Likewise, it is helpful to have students graph simultaneously the left and right sides of a trigonometric equation to see that the two functions intersect only at certain points. Solving a trigonometric equation involves finding the x-coordinates of the points of intersection. The solution to the equation $3 \tan^4 x = 1 + \sec^2 x$ is shown in figure 14.4. The functions $y = 3 \tan^4 x$ and $y = 1 + \sec^2 x$ are graphed simultaneously so that the points of intersection can be seen.

Students may also find roots of trigonometric equations graphically by transforming the equation so that one side is set equal to zero. It is interesting to present this technique to students who have previously solved trigonometric equations. Many have made no connection between the equation to be solved and its graph. Once they see that the points where the graph crosses the x-axis correspond to the roots of the equation, they can predict from the graph how many roots they are seeking and approximate values for those roots. In figure 14.5, the equation $3 \tan^4 x = 1 + \sec^2 x$ is rewritten as $3 \tan^4 x - 1 - \sec^2 x = 0$ and graphed as $y = 3 \tan^4 x - 1 - \sec^2 x$. Without zooming in at this scale, the roots can be read to one decimal place as $x = 0.8, 2.3, 3.9$, and 5.5 on a graphing calculator.

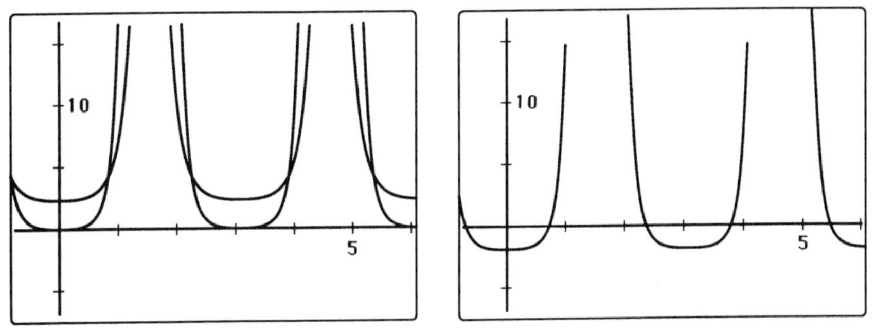

Fig. 14.4 Fig. 14.5

GLOBAL CHARACTERISTICS OF FUNCTIONS

The study of functions is rich in complex concepts. Leinhardt, Zaslavsky, and Stein (1990) review almost 200 papers on functions, graphs, and graphing. They note that students who are introduced to graphing through tabulation, plotting, reading values, sketching shapes, matching functions of the same family, identifying formulas, and finally curve sketching and curve fitting have a point-to-point focus that causes them to overlook the global characteristics of functions. Teachers would like their students to recognize the shape of a graph of a function. It is a helpful foreshadowing of calculus concepts when students generalize about the degree of a polynomial function and the number of "wiggles" or changes from increasing to decreasing it has. Maximum and minimum values seen as "peaks and valleys" on a graph are important observations.

Students can explore the domain and range needed to give a complete view of the interesting features of a function. For example, to graph

$$f(x) = \frac{x^3 - 10x^2 + x + 50}{x - 2}$$

with $x \in [-10, 10]$ and $y \in [-100, 100]$, as shown in figure 14.6, makes it apparent that a vertical asymptote can be drawn at $x = 2$ as indicated by the denominator of the formula. But, interestingly, the function looks quadratic in figure 14.7 with $x \in [-100, 100]$ and $y \in [-10\,000, 10\,000]$. This can be explained by having students perform long division to see that the function can also be written as

$$f(x) = x^2 - 8x - 15 + \frac{20}{x - 2}.$$

Graphical insight into such rational functions forms a foundation for discussions in calculus of the limit as x approaches 2 from the left and right (fig. 14.6) and the limit as x approaches positive or negative infinity (fig. 14.7). As graphing calculators make more graphs available to students, teachers need to be alert to drawing students' attention to such global characteristics.

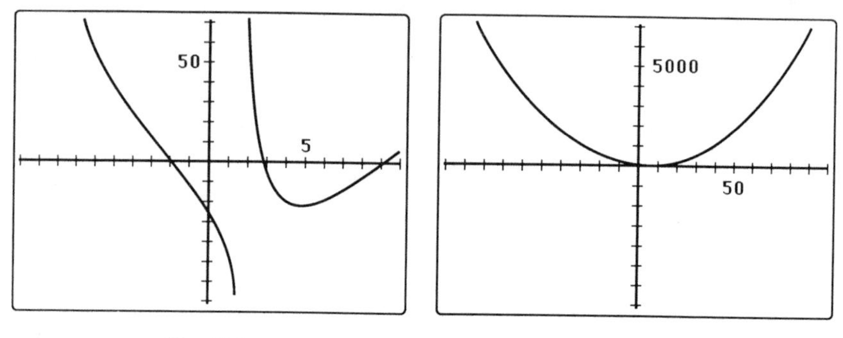

Fig. 14.6 Fig. 14.7

MISINTERPRETING GRAPHS

In using a graphing calculator, students must be cautioned to interpret results carefully. At the heart of this caution is the issue of scale and what changes and what stays the same as scale is changed. Scale was not much of an issue in mathematics instruction prior to the introduction of graphing calculators. In science, students gather data, form tables and ordered pairs, and decide on scale in the context of the units of the experiment as part of understanding what the data represent. Algebraic function rules may or may not be developed. But in mathematics instruction, problems have usually been written to have function rules simple enough for students to manipulate. Mathematical function problems have traditionally been set up to be graphed for $x \in [-10, 10]$ and $y \in [-10, 10]$. The units that generate scaling issues in function applications have been simplified so that students focus on mathematical technique rather than on science concepts. It has been especially common to keep x and y scales the same for mathematical functions. Therefore, students generalize that the graph of $f(x) = x$ is a line that bisects the first quadrant. However, when the scales on a graph are different for the x- and y-axes, the angle that the line $f(x) = x$ makes with the x-axis does not remain 45 degrees. Likewise, the angle formed by two perpendicular lines will look like a right angle only if the x and y scales are the same.

One benefit of graphing calculators is that students constantly work with scale when they search for complete graphs. They see that parabolas can be made "fat" or "skinny" by varying the x scale. Students must learn not to accept a nearly straight-line portion of a parabola as the complete graph. The quadratic algebraic representation must inform them that they are looking for more. Students need to recognize that a graph cannot be interpreted unless the scale is known. Teachers must ask students to note scale as part of their interpretation, especially when using an automatic zoom-in or zoom-out process.

A second area of concern has to do with accuracy and resolution. Graphing calculators operate with rows and columns of pixels, or picture elements. The number of x-values substituted in $f(x)$ to determine points to be plotted is limited to the number of columns. The change in x represented by each column is computed as (Xmax-Xmin)/(# of columns). Thus it is often not possible on most graphing calculators to "see" points of discontinuity. For example, the missing point in a rational function such as $f(x) = (x - 5)/(x^2 - 25)$ would not be visible. Depending on the scale, points to the left and right of a vertical asymptote of a rational function may be connected, thus giving an impression of continuity where there is discontinuity.

Algebraic techniques applied to textbook problems often yield exact answers. Graphical techniques frequently yield approximate answers, and textbook problems give little context about how accurate an approximation should be. These issues are not addressed well in many current textbooks, and teachers need to be as alert to students' graphical misconceptions as they are to algebraic misconceptions.

CONCLUSION

Graphing calculators are available and inexpensive enough to allow students to have access to them for learning about elementary functions. Their use enables students to construct many more graphs for observation and generalization than they would usually do by hand. Instruction that connects graphical, algebraic, and tabular representations of functions helps students develop richer insight into the nature of functions.

REFERENCE

Leinhardt, Gaea, Orit Zaslavsky, and Mary Kay Stein. "Functions, Graphs, and Graphing: Tasks, Learning, and Teaching." *Review of Educational Research* 60 (1990): 1–64.

15

The Power of Parametric Representations

Gregory D. Foley

PARAMETRIC equations are a powerful and flexible tool for representing curves and motion simulations in the Cartesian plane. Lines, functions, inverses of functions, conic sections, and polar curves can all be represented by parametric equations. Parametric representations can illuminate the relationships between rotation and angle measure, between vectors and their components, and between the geometry of curves and the motion of objects.

In the past, constructing the graph of a pair of parametric equations was a laborious task. Now hand-held graphing calculators (e.g., the Texas Instruments TI-81) have built-in parametric graphing utilities that automate the curve-construction process. They can simultaneously plot related curves and have a user-controlled Trace that displays a numerical readout of the parameter value and the coordinates associated with each plotted point. Graphing calculators reveal the dynamic nature of parametric representations. They can simulate the flight of a projectile or the path swept out by a point on a rolling wheel. The plotting speed can be adjusted to give the effect of a slow-motion instant replay.

Students and teachers can use this technology to gain relatively easy and early access to the interesting and useful mathematics of parametric equations and their graphs. Thus the graphing calculator can revitalize a curricular strand that enhances visualization and has applications in the mathematical, physical, and engineering sciences.

PARAMETRIC REPRESENTATIONS OF LINE SEGMENTS

Parametric equations can be used to represent any line in the plane: horizontal, vertical, or oblique. This provides a nice entry point into the mathematics of parametric equations and their graphs. Many textbooks begin their presentation of parametric representations of lines and line segments with the following sort of statement: A line that passes through the point $P = (x_0, y_0)$ and is parallel to the (direction) vector $\mathbf{v} = (a, b)$ can be represented by the parametric equations

$$x = x_0 + at \text{ and } y = y_0 + bt.$$

This is then followed by one or more examples and a set of exercises.

In contrast to this typical pattern of presentation, a guided-discovery laboratory approach is outlined below. Students can work in pairs on the series of activities, with one student using a graphing calculator and the other recording results on graph paper. Each student should be encouraged to share observations and thoughts with the other.

Line-Segment Laboratory

The goal of the graphing-calculator experiments that follow is for students to explore parametric equations of the form

$$x = x_0 + at \text{ and } y = y_0 + bt.$$

Each student is expected to prepare a report in much the same manner as is done in science classes. Students are to describe their results, search for patterns, write down conjectures, attempt to verify those conjectures or find counterexamples, and ultimately share their findings and engage in a whole-class discussion. This should lead to the class reaching, or at least approaching, a conclusion equivalent to the typical textbook rule stated above. In addition, the lab activities are designed so that students will focus on the relationship between the slope of a line segment and the coefficients of t in the segment's parametric representation.

1. Give each pair of students a single set of equations to explore, using values for x_0 and y_0 chosen from $\{-3, -2, -1, 0, 1, 2, 3\}$. Assign each group a different pair of equations, but have all groups use the interval $0 \le t \le 4$. Have them (*a*) graph on the calculator and (*b*) transfer to graph paper the curves obtained by substituting -1, 0, and 1 for a and b in all possible ways. For example, one group might explore the equations

$$x = 1 + at \text{ and } y = -2 + bt.$$

Figure 15.1 shows one possible version of the graph the students could obtain for the line segment corresponding to $a = 1$ and $b = -1$, that is, for the equations

$$x = 1 + t \text{ and } y = -2 - t.$$

The Trace feature allows the students to move a cursor along the curve and obtain a numerical readout for each plotted point. In this example the point $(3.5, -4.5)$ was plotted for $t = 2.5$.

The nine graphs the students should ultimately obtain are the eight line segments emanating from the point (1, -2) that are shown in figure 15.2 together with the point (1, -2) itself, which is a degenerate line segment corresponding to the equations

$$x = 1 + 0t \text{ and } y = -2 + 0t.$$

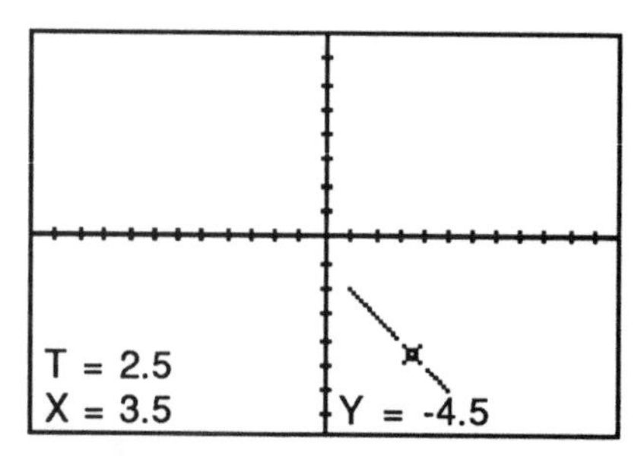

Fig. 15.1. The graph of $x = 1 + t$ and $y = -2 - t$, $0 \leq t \leq 4$, in the viewing rectangle $[-12, 12] \times [-8, 8]$, with the Trace feature displaying the coordinates (3.5, -4.5) for the parameter value $t = 2.5$

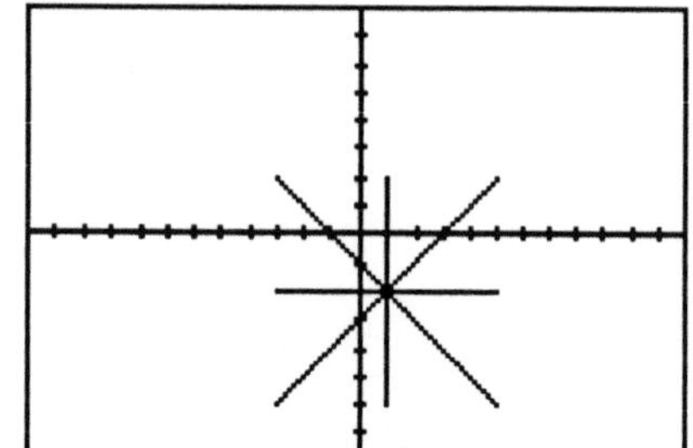

Fig. 15.2. The graphs of $x = 1 + at$ and $y = -2 + bt$, $0 \leq t \leq 4$, in the viewing rectangle $[-12, 12] \times [-8, 8]$, for all possible a, b in $\{-1, 0, 1\}$

Ask the students to specify the slope of each of the line segments they obtain. How is each slope related to the values of a and b? Have students write their conjectures.

2. Have the student teams repeat Activity 1 using the interval $-4 \leq t \leq 0$. What is the relationship between these new line segments and those obtained in Activity 1? Do the conjectures about slopes still hold true?

3. Have the teams repeat Activities 1 and 2 using other values of a and b. What are the slopes of these new segments? Do the earlier conjectures about slopes still hold, or must they be revised? Do any pairs of segments appear to be perpendicular? If so, how are their equations related? How are their slopes related?

4. Have each student, not just each pair of students, prepare a report that describes the team's results and that states and proves any definitive conclusions.

5. Have students share their findings, and lead the class in a discussion of the roles that (*a*) x_0 and y_0, (*b*) a and b, and (*c*) the interval of t values play in determining the graph. Discuss the relationship between the slope of the segments and the values of a and b in their equations. Discuss the slopes of parallel and perpendicular line segments. To move from a discussion of line segments to a discussion of lines, ask: What would the graph be if t were allowed to range over all real numbers? Then explore how the parametric equations of a line are related to its coordinate equation, $y = mx + b$.

Depending on the background and sophistication of the students and the amount of time you wish to spend on related ideas, you can extend the investigation in several directions as discussed in the next two sections.

THE GENERALITY OF PARAMETRIC REPRESENTATIONS

A wide variety of plane curves can be represented parametrically. If $f(t)$ and $g(t)$ are any two functions defined on an interval of t values, we can consider the parametric equations they yield:

$$x = f(t) \text{ and } y = g(t)$$

Their graph consists of all points $(f(t), g(t))$ for t in the given interval.

Functions and Their Inverse Relations

Any function $y = f(x)$ can be represented parametrically by

$$x = t \text{ and } y = f(t).$$

Then the graph of the inverse relation is represented by

$$x = f(t) \text{ and } y = t.$$

When this fact is used, relations like $x = y^2$ and $x = \sin y$ can easily be graphed on any calculator with a parametric graphing utility (see figs. 15.3 and 15.4). Students can be asked to determine the restrictions necessary to make the inverse relations satisfy the vertical line test for (single-valued) functions.

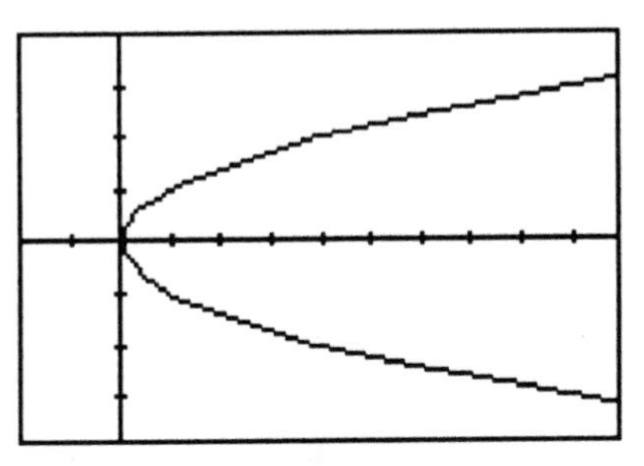

Fig. 15.3. The graph of $x = t^2$ and $y = t$, $-4 \le t \le 4$, in the viewing rectangle $[-2, 10] \times [-4, 4]$

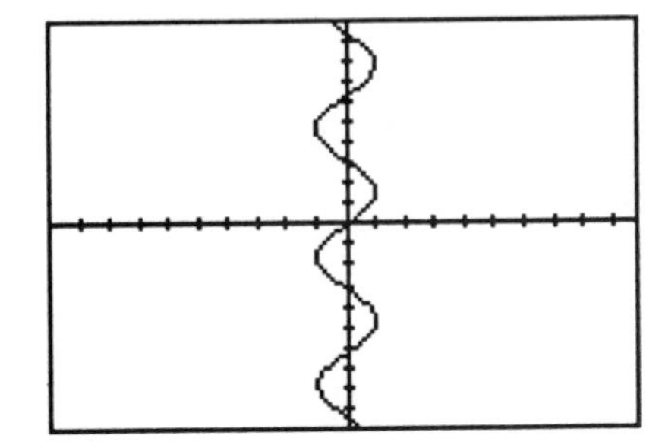

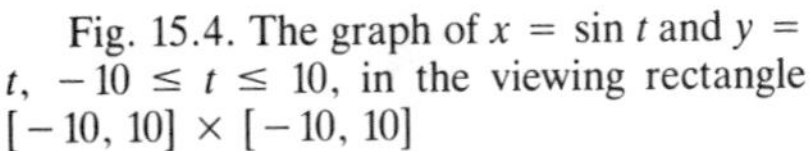

Fig. 15.4. The graph of $x = \sin t$ and $y = t$, $-10 \le t \le 10$, in the viewing rectangle $[-10, 10] \times [-10, 10]$

Graphs of functions and their inverse relations can be overlaid. For example, in figure 15.5, $y = 2^x$ and $x = 2^y$ are displayed in the same viewing rectangle. Students can experiment with several examples and be asked to explore such questions as the following: What is the geometric relationship between a function and its inverse relation? Under what conditions is the inverse of a function itself a function? To aid visualization, the graphs can be drawn simultaneously rather than in sequence, and the graph of $y = x$ (i.e., $x = t$ and $y = t$) can be overlaid (see fig. 15.6).

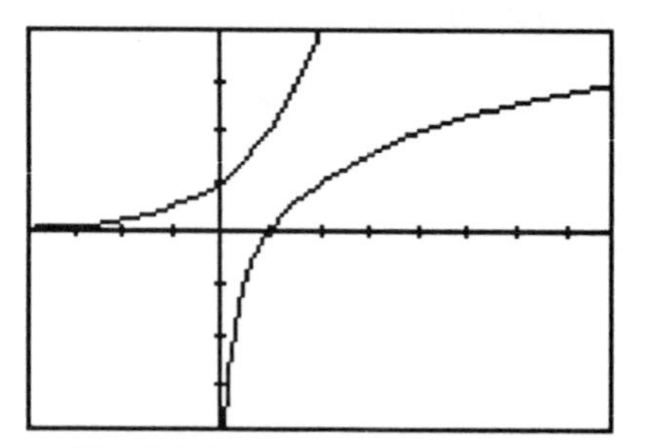

Fig. 15.5. The graphs of $x = t$ and $y = 2^t$, and $x = 2^t$ and $y = t$, $-4 \leq t \leq 8$, in the viewing rectangle $[-4, 8] \times [-4, 4]$

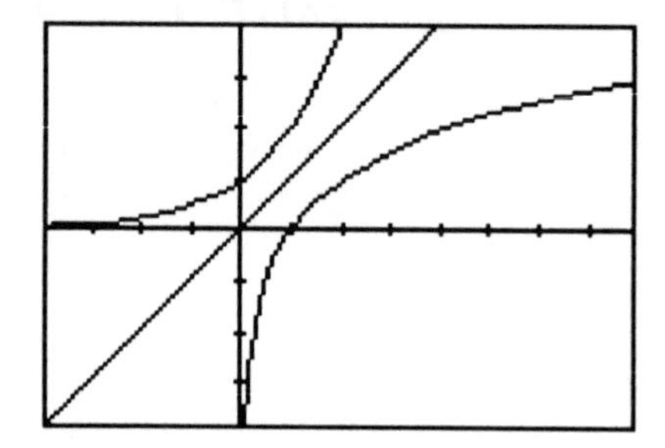

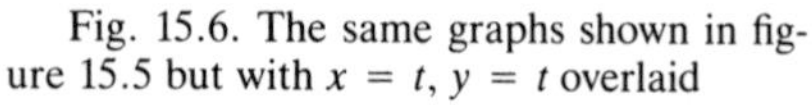

Fig. 15.6. The same graphs shown in figure 15.5 but with $x = t$, $y = t$ overlaid

Polar-Coordinate Graphs

The graph of a polar-coordinate equation of the form $r = f(\theta)$ can be represented by the parametric equations

$$x = f(t) \cos t \text{ and } y = f(t) \sin t.$$

This, together with a graphing calculator, automates the once time-consuming and tedious process of polar graphing. For example, figure 15.7 shows the graph of the three-leafed rose $r = \cos 3\,\theta$ generated by plotting the equations

$$x = \cos 3t \cos t \text{ and } y = \cos 3t \sin t,\ 0 \leq t \leq \pi.$$

Some graphing calculators (e.g., the TI-81) give the option of choosing a polar readout of coordinate values while using the Trace feature to ease the numerical exploration of polar graphs (see fig. 15.8).

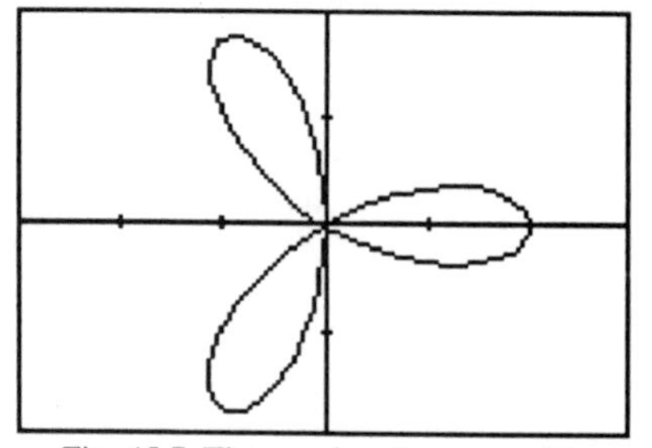

Fig. 15.7. The graphs of $x = \cos 3t \cos t$ and $y = \cos 3t \sin t$, $0 \leq t \leq \pi$, in the viewing rectangle $[-1.5, 1.5] \times [-1, 1]$

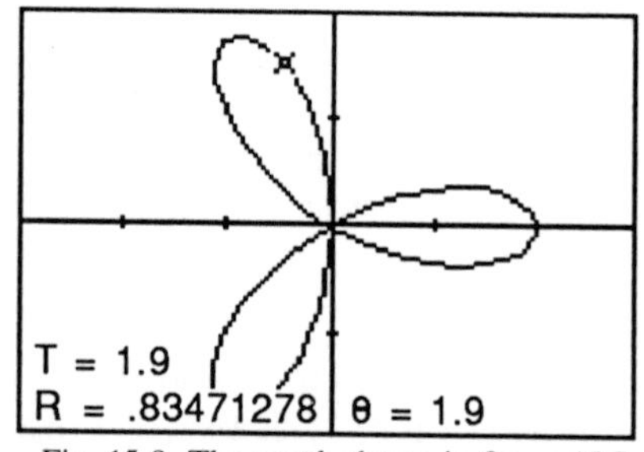

Fig. 15.8. The graph shown in figure 15.7 but with the Trace feature activated

MOTION SIMULATIONS

Thinking of the parameter t as time allows us to represent and solve a wide variety of motion problems using parametric equations. For example, simple linear acceleration can be represented nicely using parametric methods. A classic experiment in physics involves photographing an object in free fall. A strobe light is attached to the object, and the increasingly greater spacing of the light-flash images in the resulting photograph shows the object's acceleration.

Using a parametric graphing utility, we can go beyond what is possible in the physics laboratory. We can simulate and compare free-fall motion on different planets—for instance, Earth, Mars, and Jupiter. Figure 15.9 shows a simulation of the strobe-light experiment conducted on each of these planets with strobe lights being dropped from a height of 20 meters in each case, with $x = 1$ representing Earth, $x = 2$ representing Mars, $x = 3$ representing Jupiter, and the three strobes flashing every 0.1 second. In each case, $y = 20 - 0.5gt^2$, with g varying from planet to planet. (The g values, $g_E = 9.8$ m/s^2, $g_M = 3.72$ m/s^2, $g_J = 22.88$ m/s^2, come from Finney and Thomas [1990].) Figure 15.10 highlights the position of the strobe on each planet one second after being dropped, to suggest the effect obtained by moving from one graph to the next with the Trace feature activated on the screen of a graphing calculator.

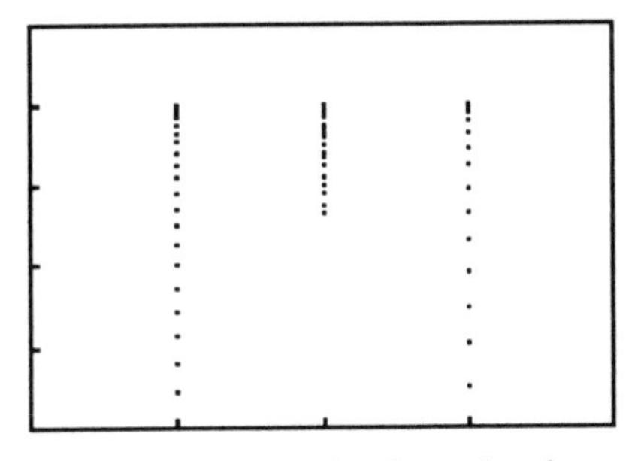

Fig. 15.9. The graphs of $x = 1$ and $y = 20 - 4.9t^2$, $x = 2$ and $y = 20 - 1.86t^2$, $x = 3$ and $y = 20 - 11.44t^2$, for $t = 0.0, 0.1, 0.2, 0.3, \ldots, 2.0$, in the viewing rectangle $[0, 4] \times [0, 25]$

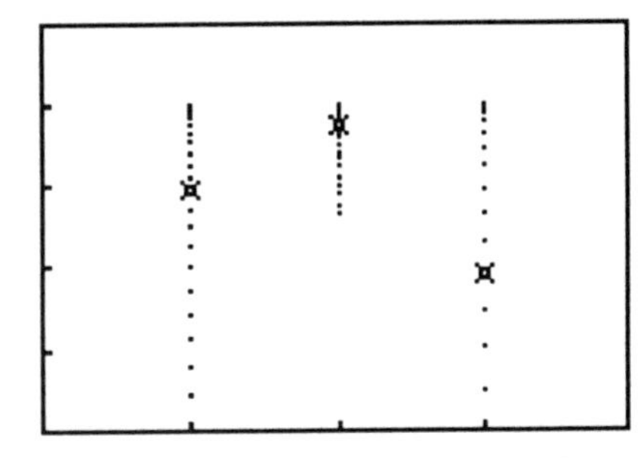

Fig. 15.10. The same graphs as in figure 15.9 but with the points corresponding to $t = 1$ highlighted

Figures 15.9 and 15.10 do not do justice to the dynamic picture produced on the calculator screen. You are urged to try this experiment on the TI-81 using the Dot and Simultaneous options on the Mode menu. It is as if three 20-meter-drop experiments on three different planets are being conducted side by side at the same time. To get the effect of a slow-motion instant replay, change the t increment from 0.1 seconds to 0.01 seconds in order to slow the plotting speed by a factor of 10.

Other examples of plane motion can be found in calculus and analytic geometry textbooks. An excellent source of examples, available in many libraries, is the "Curves and Surfaces" section of the *CRC Standard Mathematical Tables* (Eves 1969, or any other edition).

CLOSING THOUGHTS

Thorpe (1989) states that "there are three criteria, at least one of which must be met before any given topic merits inclusion in the curriculum: intrinsic value, pedagogical value, and intrinsic excitement or beauty" (p. 12). This article has attempted to show that the topic of parametric curves can satisfy all three of these criteria if presented in an appropriate manner.

"Computers and calculators change what is feasible and what is important. They make the difficult easy and the infeasible possible" (Mathematical Sciences Education Board 1990, p. 20). This is certainly true for parametric equations. In the past this powerful, general method of representation was given short shrift in schools because of the difficulty of obtaining graphs by paper-and-pencil methods. Now we can exploit the power of parametric representations without the burden of hand graphing. We should take full advantage of this new power to represent and solve problems and to explore concepts and relationships.

REFERENCES

Demana, Franklin, Bert K. Waits, Charles Vonder Embse, and Gregory D. Foley. *Graphing Calculator and Computer Graphing Laboratory Manual.* 2d ed. [Accompanies the Demana and Waits Precalculus Series.] Reading, Mass.: Addison-Wesley, 1991.

Eckert, Paul, George Kitchen, Cameron Nichols, and Charles Vonder Embse. *Graphing Calculators in the Secondary Mathematics Classroom.* Monograph No. 21. Lansing, Mich.: Michigan Council of Teachers of Mathematics, 1989.

Eves, Howard. "Curves and Surfaces." In *CRC Standard Mathematical Tables,* 17th ed., edited by S. Selby, pp. 379–94. Cleveland, Ohio: Chemical Rubber Co., 1969.

Finney, R. L., and George B. Thomas, Jr. *Calculus.* Reading, Mass.: Addison-Wesley, 1990.

Mathematical Sciences Education Board, National Research Council. *Reshaping School Mathematics: A Philosophy and Framework for Curriculum.* Washington, D.C.: National Academy Press, 1990.

Pence, Dennis. *Calculus Activities for Graphic Calculators.* Boston: PWS-Kent, 1990.

Thorpe, John A. "Algebra: What Should We Teach and How Should We Teach It?" In *Research Issues in the Teaching and Learning of Algebra,* edited by Sigrid Wagner and Carolyn Kieran, pp. 11–24. Reston, Va.: National Council of Teachers of Mathematics; Hillsdale, N.J.: Lawrence Erlbaum Associates, 1989.

16

Super Calculators: Implications for the Calculus Curriculum, Instruction, and Assessment

Thomas P. Dick

THE calculus revitalization movement has often been characterized as a move toward a "lean and lively" calculus, and the continuing evolution of technological tools like super calculators has had much to do with the reexamination of both content emphases and instructional delivery (Douglas 1986; Steen 1988; Tucker 1990). By a super calculator, we mean a hand-held *computer* with the built-in capabilities to graph functions, to manipulate symbolic expressions, including symbolic differentiation and integration, to compute with matrices and vectors, to perform high-precision numerical integration, and to find roots of functions. The access to such capabilities in a hand-held device has been encouraged for some time (see Tucker [1987]). In reporting and summarizing the technology aspects of the Sixth International Congress on Mathematical Education, Shumway (1988) argued that the recent emergence of super calculators has broad implications for the mathematics curriculum and for teaching strategies:

- *Calculators must be required* for all teaching, homework, and testing in mathematics.
- Substantial changes and redirection of the curriculum must be made to *de-emphasize numerical and symbolic computation* and *emphasize earlier, deeper, conceptual learning.*
- Teaching strategies must *de-emphasize drill and practice* and *focus on examples, nonexamples, and proofs.*

In this article we discuss the impact super calculators have made on the calculus curriculum, instruction, and assessment in the Oregon State University Calculus Project, in which every student has constant access to either an HP-28S or an HP-48SX super calculator. We have found that super calculators can fundamentally redefine the teacher and student environment, but their use does introduce some pitfalls.

ONE STUDENT'S USE OF A SUPER CALCULATOR

Ronnie is a (real) student in a college calculus class using super calculators. There are times when Ronnie might be described as symbolically *dysfunctional.* When faced with a purely symbolic manipulation task, Ronnie appears to fall back on a collection of disconnected and incomplete rote-memorized rules. For example, if Ronnie had been given the exercise of evaluating the definite integral

$$\int_0^{\ln 2} (e^x - e^{-x})dx$$

symbolically *by hand,* the multiple opportunities for incorrect signs and other symbolic errors (such as $e^{-\ln 2} = -2$) made it quite likely that Ronnie would fail to obtain the correct answer. Although a case could be made for the use of super calculators simply to help students like Ronnie with basic calculus computations, the greater promise of these machines lies in their use as a problem-solving aid for all students.

Consider the following problem: Find the area of the region bounded by the graphs of $y = e^x$, $y = e^{-x}$, and $x = \ln 2$. The "usual" solution to this problem would be to compute

$$\int_0^{\ln 2} (e^x - e^{-x})dx = (e^x + e^{-x})|_{x=\ln 2} - (e^x + e^{-x})|_{x=0}$$

$$= (2 + \frac{1}{2}) - (1 + 1) = \frac{1}{2}.$$

In contrast, Ronnie's solution to the problem was to compute

$$[\ln 2 - \int_1^2 (\ln x)\, dx] + [\frac{1}{2}\ln 2 - \int_{1/2}^1 (-\ln x)\, dx]$$

$$\approx [0.693144718 - 0.386294361] + [0.346573590 - 0.153426410] \approx 0.5.$$

At first glance, the instructor was perplexed and assumed that this could be a case of "a right answer for the wrong reason." After further study, however, it became clear that Ronnie had on the super calculator the two functions $y = e^x$ and $y = e^{-x}$, and his attention was drawn to the intersection point (0, 1) (see fig. 16.1).

Conceivably, Ronnie noted the area as breaking down into two components above and below the line $y = 1$. He determined the y-coordinates of

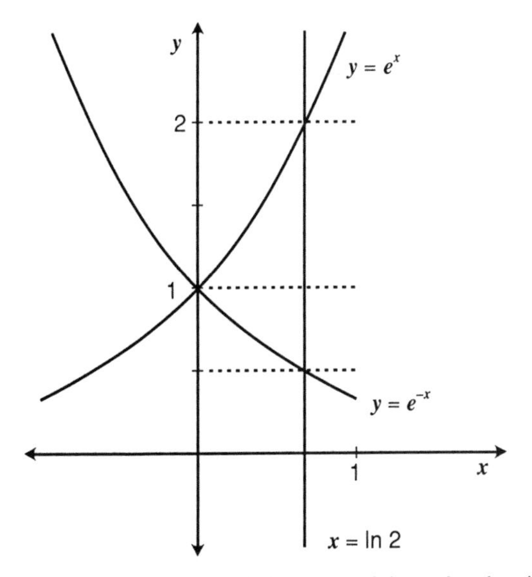

Fig. 16.1. Ronnie's problem: Find the area of the enclosed region

the intersections of the two graphs with the line $x = \ln 2$, that is, $y = 1/2$, 2. He proceeded through subtasks to find the desired area:

1. Find the area between the graphs of $y = e^x$, the y-axis, and $y = 2$.
2. Subtract this from the area of the rectangle bounded by $x = 0$, $x = \ln 2$, $y = 1$, and $y = 2$.
3. Find the area between the graphs of $y = e^{-x}$, the y-axis, and $y = 1$.
4. Subtract this from the area of the rectangle bounded by $x = 0$, $x = \ln 2$, $y = 1$, and $y = 1/2$.
5. Add the results of steps 2 and 4.

Essentially, Ronnie rewrote the functional equations in terms of y instead of x in the integrals for steps 1 and 3, and used x as the variable of integration. Ronnie numerically integrated the definite integrals on his super calculator, computed the areas of the rectangles (by paper and pencil), and then completed his work. In looking at the graph, Ronnie judged his answer of 1/2 to be reasonable.

The super calculator *empowered* Ronnie with graphic and numeric tools, and he exploited those tools. Did Ronnie solve the problem the "hard way"? For those schooled in traditional methods, perhaps so; but for students with super calculators, I think not. Ronnie devised a logical and straightforward solution with *his* visual perception of the area's components, and he moved easily among graphical, numerical, and symbolic representations in carrying out his plan.

It is also striking that Ronnie approached the problem in the spirit of Polya and not by the traditional "match and apply" strategy of finding the right symbolic template in a textbook. Certainly, Ronnie's use of the super calculator was not a simple application of a black box; Ronnie needed a firm grasp of the relationship between integral and area to understand the problem and devise his plan of solution. The definite integrals Ronnie set up would require integration by parts if computed symbolically (an integration technique not yet covered in the course), but with the numerical tools in hand, Ronnie could forge ahead. Finally, Ronnie looked back at the graph as a monitor of the correctness of his result, and he did not look back at the end of the book for a solution.

This example points out two important changes that super calculators can bring to the mathematics classroom:

1. A greater access to multiple representations of functions
2. A greater opportunity for a problem-solving focus

ORGANIZING THE CALCULUS CURRICULUM, INSTRUCTION, AND ASSESSMENT AROUND MULTIPLE REPRESENTATIONS

Since the concept of function is central to mathematics, curricular reform at both the secondary school and the college levels must reflect an emphasis on multiple representations in the study of the concept. Improving students' abilities to move among tabular, symbolic, and graphical representations of functions are explicit objectives in the NCTM *Curriculum and Evaluation Standards* (National Council of Teachers of Mathematics 1989). In Israel, a "Triple Representation Model" (TRM) curriculum is being developed for teaching the function concept (Schwarz, Dreyfus, and Bruckheimer 1990), and many of the current calculus projects place an increased emphasis on multiple representations of functions. For example, the Harvard calculus project uses the term "Rule of Three" to refer to the numeric, symbolic, and graphical approaches to problems.

Lack of access has been the overriding obstacle to the routine use of numerical and graphical representations in instruction and assessment. Performing lengthy numerical computations and plotting graphs "by hand" are so time consuming that reliance on symbolic methods is often de facto the most efficient approach. By greatly improving access to numerical and graphical representations, super calculators make it reasonable for students to use numerical and graphical methods. They appear to be devices made to order for a calculus curriculum that emphasizes multiple representations.

Appeals to numeric and graphical representations are not new revelations in teaching the function concept. Tall and Vinner (1981) point out that many teachers introduce the definition of function with a range of representations and then spend long periods of time with only symbolic examples. The same can be said for many topics in calculus. After motivating our students with graphs and numbers, we tend to get down to the "real" work of juggling symbols. Furthermore, traditional assessment has been dominated by testing skills in manipulating symbolic expressions and very rarely done by testing understandings based on other representations. With this experience, students create a personal concept of function that involves only the symbolic setting, and they leave the course with an impoverished view of function as merely a symbolic expression to manipulate.

If we wish students to make *regular* use of numeric and graphical representations as aids to understanding the behavior of functions, as heuristics in problem solving, and as monitors of the correctness of their symbolic analysis, then they need to view these approaches as both natural and worthwhile. Viewing an activity as natural does not come "naturally"—it comes from seeing it as an everyday occurrence. The instructor's significance as a role model is key in this respect, particularly in discussing and working through problem examples. Although an instructor may not cover as many examples by regularly taking time to appeal to graphs and numerical values, this activity better equips students to handle new problems and variations.

Bridges are not built by accident. If students are to build appropriate mental connections between representations, then classroom and homework activities must be designed to guide students to insights into the power of representational connections. For example, the use of the super calculator in making numerical approximations need not be limited to such topics as Newton's method or numerical integration. Indeed, a strong case can be made for numerical approximations as a primary tool in investigating and studying limits, continuity, and differentiation. Deltas and epsilons have long been under attack as unnecessarily rigorous trappings for a first course in calculus. I would maintain that the language of precision and estimation gives natural and meaningful interpretations to deltas and epsilons as tolerance bounds. However, the symbolic tricks of the trade (such as canceling common factors in a difference quotient) can obscure the true nature of the limit process for many students.

Similarly, graphical techniques and explorations provide a whole new language with which to impart rich meaning to the central concepts of calculus. The notion of a differentiable function as being approximately locally linear can be seen dynamically when zooming in on the graph. This visual idea of derivative as local slope can be appealed to time and time again in studying the properties and applications of the derivative. For example, if two differentiable functions share a common zero at $x = a$, then zooming in on the graphs at that point can suggest L'Hôpital's rule in a natural way. Certainly, work with polar and parametric equations can be greatly enhanced with the availability of a grapher, and the quality of function approximations such as Taylor polynomials and cubic splines can be appreciated visually.

In assessment, one of the often asked questions is, "If we are going to make use of super calculators in the classroom, won't we need some tests that really *require* the use of super calculators?" The question is misguided, and it can lead to some silly test designs. Assessment should not be driven by the available tools, but those tools must be kept in mind when designing tests. Simply substituting decimal numbers to three places for integers is no advance in assessment, nor is computing a definite integral if it is no more than a button-punching exercise on a super calculator; however, posing a problem whose solution relies on computing a definite integral could be appropriate in the context of problem solving.

Ironically, some of the ways we assess students' use of multiple representations often do not involve the super calculator at all. For example, consider the graph in figure 16.2, and pose the following questions: "What is the graph of $y = e^{f(x)}$? What is the graph of the derivative of f? What is the graph of at least one antiderivative of f? (*Hint:* Use one of the fundamental theorems of calculus.) Where is $1/f$ differentiable?

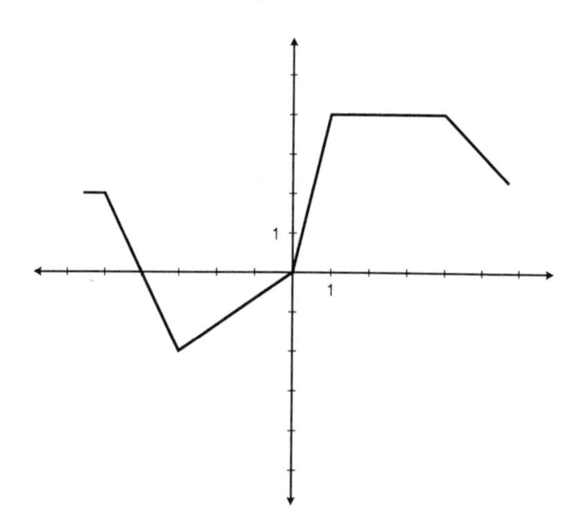

Fig. 16.2. Graph of a piecewise linear function f

These questions require students to deal explicitly with interchanges between graphical and numerical representations. The questions are not original ideas, and they can be posed in nontechnological settings. But these questions assess understanding and skills that are especially relevant to the best use of super calculators. Just as we should make a quick mental estimate of a numerical computation as a check on the reasonableness of a calculator's arithmetic result, we could and should make a quick mental sketch of the graph of a derivative as a check on the reasonableness of a super calculator's symbolic presentations.

We should not be surprised when students who spend significant time working with multiple representations on a super calculator turn out to be less proficient than traditional mathematics students on purely symbolic computations without the machine. A curriculum organized around technology and multiple representations would be inadequately assessed without noting the difference in the quality of the students' use of multiple representations. The primary skills to emphasize and to assess in a multiple representation environment are those in interpreting and translating information presented in numeric, graphic, and symbolic forms.

SUPER CALCULATORS AND PROBLEM SOLVING IN CALCULUS

There are three ways in which opportunities for problem solving are improved by the use of a super calculator:

1. More instructional time
2. More problem-solving tools
3. Student perception of problem solving

The first may be the most obvious. The need for spending large amounts of instructional time on symbolic computational skills seems questionable in the presence of a machine that allows symbolic, numeric, and graphical methods. Super calculators have already generated a "basic symbolic skills" debate at the college level that is remarkably similar to the continuing controversy over four-function calculators and basic arithmetic skills at the elementary school level (Dick 1988). Time liberated from symbolic computation could be spent on mathematical modeling and problem solving. In a study involving resequencing skills and applications instruction in a college calculus course with computer software, Heid (1988) gave evidence that it is possible to substantially and successfully compress the time spent on symbolic manipulation skills and expand the time spent on applications and problem solving. In our own super calculator calculus project, we have been able to double the class time we spend on applications and problem solving.

By the second point, I do not mean simply the use of the super calculator for carrying out routine computations, but its use as a heuristic and monitoring aid. With greater access to numeric and graphical representations comes the opportunity for students to try special numerical cases easily and to graph as a first step instead of a last one. Reconciling symbolic computations against numeric and graphical data is a powerful way to "look back."

Third, the availability of a super calculator has an affective consequence in problem solving. With powerful numeric, graphical, and symbolic computational tools in hand, the student can see the "carry out the plan" stage of problem solving as the *least* daunting step. Students can appreciate more the relative importance of heuristic processes, mathematical modeling, and the interpretation of results. In contrast, when symbolic computation by hand is the only means available, its importance is exaggerated in the student's perception, and students can come to view the initial steps of understanding a problem situation and mathematically modeling it correctly as the difficult but minor "setting up" phase. Problem-solving researchers should take a special look at the classroom equipped with super calculators, for it is a really *new* environment unburdened by some of the baggage of the traditional classroom.

NEW SKILLS NEEDED TO USE SUPER CALCULATORS

Super calculators give students greater access to information from multiple representations. The new skills required to use super calculators effectively can be categorized according to the limitations and constraints the technology places on the quality of this information.

Numerical Skills

Using numerical approximation methods requires more attention to the precision limitations of a machine. Round-off and cancellation errors need to be considered in a sequence of calculations to judge the accuracy of a numerical result. The cumulative effects of round-off errors in numerical computations are easily misunderstood by students. Sometimes students underestimate round-off effects by mistakenly assuming they are additive. At other times, students may grossly *overestimate* round-off errors as a convenient scapegoat in explaining away conflicting numerical and graphical or symbolic information. (This can occur even when the order of magnitude makes that explanation untenable.) The important distinction between relative and absolute error is also fuzzy to many students.

Graphical Skills

Any graphical device necessarily has a finite viewing screen, and that limitation creates a need for scaling and positioning skills in the creation of the viewing window of the Cartesian plane. Graphical evidence is open to perceptual illusions and therefore to misinterpretations, particularly by inexperienced students (Goldenberg 1988). Since different graphical operations can lead to a similar visual picture (e.g., simple translation of axes gives the same picture as both changing the center of the viewing window and translating the function), attributions of cause and effect in seeing graphical changes can be quite tricky. Graphical behavior can also be hidden by—

1. lying beyond the bounds of the finite viewing window;
2. scale (zooming in obscures global information and zooming out obscures local information about the graph);
3. limited numerical precision (round-off errors are involved in the discrete selection of pixels to describe a plot).

Symbolic Skills

The importance of *context* is crucial in effectively using any symbolic algebra system. By context, we mean that the same symbols can carry different meanings and interpretations depending on when and where we use them. For example, $y = 3$ can represent simply a specific replacement value for the variable y in one context, or it might refer to a defining formula for a *constant function* in another context. The symbol dy/dx has a very important meaning in calculus that is quite different from the interpretation of it as a fraction of two algebraic quantities.

In order to manipulate symbolic expressions, a symbolic algebra system must know the context in which it is working. If we do not communicate this context effectively to the system, then it may be forced to make some assumptions that do not match our intentions. Sometimes this can result in context errors, meaning errors due to a mismatch of assumed and intended meanings for the symbols. For example, canceling the d's in dy/dx would be grossly inappropriate in the context of its usual meaning in calculus.

In short, we manipulate symbols with some meaning attached to them in our minds. Symbolic algebra systems manipulate symbols according to certain rote procedures and patterns. To use a symbolic algebra system intelligently requires monitoring closely the match between the context the system uses and the meanings we intend for the symbols. Unleashing the symbolic power of a super calculator demands communicating with the machine in an unambiguous language. This means that students must pay careful attention to the meanings and interpretations of the symbols that they use.

Translation Skills

Besides those skills needed for using a particular representation are those related to moving *between* representations. Estimating the numerical value of a definite integral from a graph or judging the reasonableness of a symbolic limit computation from numerical evidence are examples of useful multiple-representation skills in calculus, particularly when a super calculator is available. However, we must be wary of students' willingness to make very strong conclusions on the basis of incomplete graphical or numerical evidence. A "proof by picture" becomes quite tenuous when the picture represents a finite collection of ordered pairs generated by a machine. Although graphical and numerical results of limited precision can *suggest* much to us, much thoughtful analysis is needed before making mathematical conclusions.

Confrontational Examples

In our calculus classes using super calculators, we often use *confrontational examples*—that is, examples chosen to confront the limitations of the super calculator or what we have found to be common misconceptions. For example, the "hole" in the graph of a function with a removable discontinuity may show up—

1. as a missing pixel (occurs if a column of pixels has an abscissa at the hole);
2. as a jagged "jump" or "drop" (may occur if a column of pixels has an abscissa sufficiently close to the hole);
3. not at all (the graph appears to be "continuous").

By having students change the viewing window slightly, we can make sure they experience each one of these phenomena and explain why it happens. As another example, repeatedly zooming in on the graph of a continuous function can lead to chaotically displayed pixels, which are due to numerical round-off errors in computing pixel coordinates from column to column. By drawing students' attention to such limitations explicitly, we hope to breed a healthy skepticism that can ward off erroneous conclusions. Much research needs to be done in describing subtle but problematic misconceptions that may arise from students' increased use of numeric and graphical representations.

RESEARCH QUESTIONS

Some mathematics education research questions have particular relevance to the calculus classroom equipped with super calculators. Studies of these questions could help us determine the most effective pedagogical use of super calculators.

Cognitive Issues

• *What are the cognitive obstacles to translating from one representation to another and how can they be overcome?* Once the use of a particular representation has been "triggered," students may find it difficult to transfer to a different representation. This phenomenon is often called the "compartmentalization" of function representations (Markovits, Eylon, and Bruckheimer 1986; Vinner and Dreyfus 1989). Even though super calculators give greater access to numeric and graphical representations, we still see this phenomenon of compartmentalization, and of course it adversely limits the potential uses of the super calculator. How do we "decompartmentalize" students' notion of function, and in particular, can a super calculator be exploited in special ways to accomplish this?

• *When a super calculator is available, how do students "manage" the multiple representations?* When a student gets conflicting information from multiple representations, how is the conflict resolved? Is there a representation that students "trust" more than others, and why? Does it make sense to classify students according to their use of, or preference for, representations? We have certainly noted some differences in the frequency with which students make use of graphical and numerical tools when undirected. What characteristics (of tasks, of students) affect a student's choice of representation?

• *Does the use of an external source of representational information enhance or deter the internalization of representational knowledge?* Super calculators bring a new, *external* source of access to numeric and graphical representations and new opportunities for translating between representations. But the richness of one's personal concept of function depends in large measure on the quality of *internal* access to multiple representations. That is, regardless of the availability of technology, we wish for students to build a mental network made up of easily retrievable and interconnected symbolic, numeric, and graphical examples, counterexamples, and interpretations. Is it possible that easy external access might deter students from creating for themselves a ready *internal* access to multiple representations of functions? Or can we find evidence that repeated use of externally produced representational information leads to internalization? If not, are there instructional activities that can promote this internalization?

• *How will the role of symbolic computation change?* Just how much time can be "safely" liberated from symbolic manipulation instruction? Heid's study suggests that it might be substantial. How much and what type of symbolic manipulation skill development is "enough" for students to make effective use of symbolic representations and to translate among symbolic, numeric, and graphical representations?

Ronnie's case is evidence that a lack of symbolic manipulation skills need not exclude one from understanding and using mathematics in a productive way. The intelligent manipulation of symbols becomes more and more important in higher levels of mathematics. Is it possible that we will do "harm" to those students who go on to advanced mathematics study if time spent on symbolic manipulation skills is reduced?

Communication between student and machine itself may bring some side benefits to symbol manipulation. Super calculators naturally demand the use of unambiguous notation. Thus, to realize the savings in time and to harness the power of computation that a mathematical tool kit can provide, students will need to pay more, not less, attention to understanding the meaning of the symbols and notation they use.

Affective Issues

• *How do students' attitudes toward mathematics and beliefs about the nature of mathematics change as a result of using super calculators?* The change in students' perception of the problem-solving process has been mentioned. Some students feel insecure about their perceived dependence on a mathematical tool kit, believing that machine use is "cheating." Is this ultimately a temporal concern? Only preliminary studies have been made on the effects of super calculator use on attitudes and beliefs about mathematics (Dick and Shaughnessy 1988). Much more could be done.

In our project we have noted students who have made changes in their academic plans (all to include more mathematics) that they attributed directly to their super calculator experience. This particular observation deserves a closer look. "Greater access to representations can lead to greater access to scientific careers" has a ring to it worth exploring.

• *How does the teacher's perceived role change in a mathematics classroom using super calculators?* A very common report from teachers making use of super calculators in the classroom has been an increase in "discovery" activity on the part of students. Some of it is guided and planned, but the use of the machines also brings about a lot of "Why did this happen?" and "What would happen if?" questions. These furnish some spontaneous, but golden, opportunities to launch mathematical discussions or student explorations, but they require more flexibility on the part of the teacher. This flexibility may be a bigger challenge for college faculty than for secondary school teachers.

CONCLUDING REMARKS

The existence of powerful mathematical capabilities in a hand-held package provides many rebuttals to the traditional reservations regarding affordability, availability, and accessibility offered in resistance to the use of technology in the calculus classroom. Calculus students and teachers at both the secondary school and the college levels are grappling with an exciting and new classroom environment. In summary, the needs in curriculum, instruction, and assessment for the calculus class equipped with super calculators are as follows:

Curriculum: Materials must promote the use of numeric and graphical representations by appealing to them again and again in examples and exposition, not just in the introduction of a new topic.

Instruction: Teachers must *model* the use of numerical and graphical representations *explicitly* and *often,* particularly as a heuristic in problem solving and as a monitor of symbolic solution analysis. At the same time, the limitations of machine-generated information must not be sidestepped.

Assessment: Students need to be routinely assigned, and tested using, tasks that require *explicit* use of numeric and graphical analysis as well as tasks that require translation among the various representations.

REFERENCES

Dick, Thomas. "The Continuing Calculator Controversy." *Arithmetic Teacher* 35 (April 1988): 37–41.

Dick, Thomas P., and J. Michael Shaughnessy. "The Influence of Symbolic/Graphic Calculators on the Perceptions of Students and Teachers toward Mathematics." In *Proceedings of the Tenth Annual Meeting of PME-NA*, pp. 327–33. DeKalb, Ill.: North American Chapter of the International Group for the Psychology of Mathematics Education, 1988.

Douglas, Ronald S., ed. *Toward a Lean and Lively Calculus.* MAA Notes, vol. 6. Washington, D.C.: Mathematical Association of America, 1986.

Goldenberg, E. Paul. "Mathematics, Metaphors, and Human Factors: Mathematical, Technical, and Pedagogical Challenges in the Educational Use of Graphical Representation of Functions." *Journal of Mathematical Behavior* 7 (1988): 135–73.

Heid, M. Kathleen. "Resequencing Skills and Concepts in Applied Calculus Using the Computer as a Tool." *Journal for Research in Mathematics Education* 19 (1988): 3–25.

Markovits, Zvia, Bat-Sheva Eylon, and Maxim Bruckheimer. "Functions Today and Yesterday." *For the Learning of Mathematics* 6 (3) (1986): 18–24.

National Council of Teachers of Mathematics. *Curriculum and Evaluation Standards for School Mathematics.* Reston, Va.: The Council, 1989.

Schwarz, Baruch, Tommy Dreyfus, and Maxim Bruckheimer. "A Model of the Function Concept in a Three-Fold Representation." *Computers and Education* 14(3)(1990): 249–62.

Shumway, Richard J. "Technology, Mathematics, and the International Congress." Special report on ICME-6, Budapest, Hungary, 1988.

Steen, Lynn A., ed. *Calculus for a New Century.* MAA Notes, vol. 8. Washington, D.C.: Mathematical Association of America, 1988.

Tall, David, and Shlomo Vinner. "Concept Image and Concept Definition in Mathematics with Particular Reference to Limits and Continuity." *Educational Studies in Mathematics* 12 (1981): 151–60.

Tucker, Thomas. "Calculators with a College Education?" *MAA Focus* 7 (1) (1987): 1, 5.

———. *Priming the Calculus Pump: Innovations and Resources.* Washington, D.C.: Mathematical Association of America, 1990.

Vinner, Shlomo, and Tommy Dreyfus. "Images and Definitions for the Concept of Function." *Journal for Research in Mathematics Education* 20 (July 1989): 356–66.

17

The Use of Calculators in the Assessment of Mathematics Achievement

Martha H. (Marty) Hopkins

THE most common reason cited by teachers for not using calculators in mathematics classrooms is that their use is not permitted on standardized tests (Wilson and Kilpatrick 1989, p. 7). A 1987 survey of technology policy in the United States and Canada conducted by the National Council of Teachers of Mathematics (NCTM) indicated that although 40 percent of states recommend using calculators for instruction, only three have mandated their use in statewide testing—all at the secondary level only (Kansky 1987). In an effort to change this situation, several organizations and school systems have issued policy statements either encouraging or mandating the use of calculators in both the teaching and the assessment of mathematics (NCTM 1987; North Carolina State Department of Public Instruction 1983; California State Department of Education 1986; National Council of Supervisors of Mathematics 1989; Florida Chamber of Commerce et al. 1989). In general, these recommendations are based on the premise that the use of calculators in assessment will (1) ensure that students do not spend a disproportionate amount of time on computation during testing; (2) allow the inclusion of realistic problems on the tests, thus making them a more reliable measure of students' conceptual understanding as opposed to their computational proficiency; and (3) promote a shift in the mathematics curriculum away from computation to problem solving and reasoning.

Reports of calculator use in assessment are evident in the recent literature. Although results are often sketchy at best, three questions seem to appear consistently: (1) What content is most appropriately tested with a calculator? (2) How do test items on calculator-based tests differ from those on tests not allowing calculator use? (3) Should calculator use be optional or required during testing?

This research was funded by the Florida Department of Education. The opinions and interpretations expressed in this article do not necessarily reflect the position, policy, or endorsement of the Florida Department of Education.

Part 4

WHAT CONTENT IS MOST APPROPRIATELY TESTED WITH A CALCULATOR?

Objectives tested with calculators generally fall into one of two categories: (1) calculator-specific, and (2) mathematics objectives. Calculator-specific test items assess students' ability to manipulate the tool. Such objectives typically begin with "The student will use a calculator to. . . ." Examples of objectives in this category appear in the *Michigan Essential Goals and Objectives for Mathematics Education* (Michigan Department of Education 1988), including (*a*) the recognition of appropriate key sequences for automatic constant features, (*b*) the recognition and interpretation of the calculator display, and (*c*) the recognition of specific calculator keys. A rationale for such objectives is offered in the article by Payne in this yearbook. Although the mastery of these objectives is a prerequisite for appropriate use of calculators, the joint symposium of the College Board and the Mathematical Association of America recommended that mathematics achievement tests should be curriculum based and that no questions should be used that measure *only* calculator skills or techniques (Kenelly 1989).

Mathematics objectives commonly cited as appropriate for testing with a calculator include (*a*) the exploration of number patterns, (*b*) the use of a guess-and-check strategy for problem solving (Wilson and Kilpatrick 1989), (*c*) the process of hypothesis formation and verification (Heid 1988), and (*d*) the solution of problems using realistic data. With one notable exception (Reys 1980), calculators have typically been prohibited when assessing basic computation skills (Oregon State Department of Education 1985; Carter and Leinwand 1987; North Carolina State Department of Public Instruction 1983).

ITEM SPECIFICATIONS FOR CALCULATOR-BASED AND NONCALCULATOR TESTS

The use of calculators in testing allows for greater flexibility in the construction of test items. The calculator may allow students to concentrate on strategic approaches to problems without getting tangled up in memorized computational algorithms. For example, a child who has not yet learned how to divide may use the calculator to solve the following problem:

> Doughnuts cost 19 cents each. If Sally has $1.35, what is the largest number of doughnuts she can buy?

Typical solutions to this problem include the use of successive subtraction, successive addition, or estimation in conjunction with a guess-and-check strategy. In any event, the student need not be able to divide 1.35 by 0.19

in order to demonstrate an understanding of the required problem-solving process.

The use of calculators also enhances the validity of certain items by permitting the use of more realistic numbers. This allows problem-solving situations to be more akin to those found in practice (Wilson and Kilpatrick 1989). For example, the following item, which is currently appropriate for intermediate-grade students, may be accessible to students in lower grades who have access to calculators:

> In servicing a car, the attendant used 5 quarts of oil at $1.65 a quart and 15 gallons of gasoline at $1.09 a gallon. What was the total cost for oil and gas?

The reasonableness of distractors and the amount of time required to complete individual items have been identified as significant issues to consider when designing test specifications for calculator-based mathematics tests (Long, Reys, and Osterlind 1989; Carpenter et al. 1981). While studying the use of calculators in testing, Long, Reys, and Osterlind (1989) found that the selection of distractors representing the most common calculator errors "often produces responses that challenge the definition of plausible" (p. 323). During the field-testing portion of their study, however, it was found that even the most outlandish answers were attractive to the students and were selected.

The issue of time is also critical. The National Assessment of Educational Progress (NAEP) studies allowed some students to use calculators on all items during the 1977–78 testing (after scant preteaching on the use of the calculator). Students using the calculators seemed to spend more time on the test items than students not allowed to use calculators. What is unclear from these studies is whether the extra time was needed because of unfamiliarity with the calculator or because of the types of problems appearing on the test. In any event, it appears likely that when students are given problems for which they have not learned an algorithm (see examples above), additional time would be needed for the guess-and-check behaviors that are likely to result or for the students to devise a plan of attack. Not providing the extra time may cause students to leave many items blank or merely choose the "don't know" option (Carpenter et al. 1981).

SHOULD CALCULATOR USE BE OPTIONAL OR REQUIRED DURING TESTING?

With the exception of the National Assessments of Educational Progress and the College Board Mathematics Level II Achievement Test, calculator assessment projects have typically allowed students the *option* of using a calculator during testing (Alberta Department of Education 1981; Dye 1981;

Oregon State Department of Education 1985; Leitzel and Waits 1989). The joint symposium of the College Board and Mathematical Association of America held in 1986 recognized that choosing whether or not to use a calculator when addressing any particular test item is itself an important skill. Consequently they recommended that calculators should not be required on all test items on a calculator-based mathematics test. Individual items can (and should) be constructed such that using a calculator may help, hurt, or not affect performance (Kenelly 1989).

In summary, it appears that previous studies related to the use of calculators in assessment suggest that (1) items included should test problem-solving rather than basic computation skills, (2) item specifications for calculator-based tests should differ from those for tests not allowing calculator use, and (3) although calculators should be available to all students during testing, their use on individual items should be optional.

States throughout the country are currently studying the implementation of assessment recommendations in an effort to provide assessment that matches more appropriately the curriculum reforms of the 1990s. The following project was completed during the 1989–90 school year by the state of Florida in an attempt to adjust its statewide assessment program accordingly.

THE FLORIDA PROJECT

The purpose of the project was to investigate the feasibility of using calculators in the assessment of Florida students' mathematics proficiency. At the writing of this report, the project had completed three phases: (1) identifying objectives to be tested, (2) writing test specifications and items for the identified objectives, and (3) pilot-testing the items. Each phase will be discussed separately.

Identifying Objectives

One hundred two (102) surveys were distributed to mathematics contact personnel throughout the state of Florida in an effort to determine which objectives in the mathematics program are most suitable for testing with the use of a calculator. All mathematics objectives appearing on the *Minimum Student Performance Standards of Florida* (Florida State Department of Education 1985) and *Student Performance Standards of Excellence for Florida Schools in Mathematics, Science, Social Studies, and Writing* (Florida State Department of Education 1984) were included on the surveys. Objectives included basic computation (Minimum Standards) and problem-solving skills (Standards of Excellence). Beside each listed objective the contact person was to check whether that objective should, could, or should not be tested with a calculator. Space was provided for contact people to mention any particular calculator they would suggest be used and to write any general comments.

Ninety-one (89.2%) of the surveys were returned and tabulated. All objectives receiving feedback favoring calculator use were identified. Close inspection of these objectives revealed the following:

1. *Elementary*—Eight objectives met the criteria at the elementary school level, four each at grades 3 and 5 where testing occurs. All eight of these objectives require the solution of real-world problems involving addition, subtraction, multiplication, or money.

2. *Middle school*—Fifteen objectives met the criteria at the eighth-grade level. Six of the objectives appear on the Standards of Excellence; the remainder are skills from the Minimum Standards. Once again, all objectives require the student to solve real-world problems involving the four basic operations, money, measurement, and geometry.

3. *High school*—Twenty-one objectives met the criteria at the twelfth-grade level. Twelve of them, however, dealt with concepts such as exponential and logarithmic functions, real and complex roots of a number, and probabilities of compound events. Although these objectives appear on the Standards of Excellence, it was determined that they are most likely not part of the curriculum of the average twelfth-grade student. Consequently, these objectives were deleted from the final list. The remaining nine objectives appear on the Minimum Standards and require the solution of real-world problems involving the four basic operations, percent, money, and measurement.

Close inspection of all data revealed that most mathematics personnel surveyed in the state of Florida felt that calculators should be used to test problem-solving skills but not basic computation. This was confirmed by results of research previously reported in the literature. Survey results also indicated that the TI-108 calculator was the preferred tool at grade levels 3, 5, and 8. Choices for calculators varied widely at the twelfth-grade level.

Writing Test Specifications and Items

On completion of the first phase of the project, item specifications were written for each of the objectives, with two overriding considerations: (1) it was assumed that students in all grades (5, 8, and 11) would have access only to a simple four-function calculator that did not have the order of operations programmed into it; and (2) all word problems would contain situations familiar to students taking the test. One of the goals of the project was to ascertain the extent to which calculator-based item specifications differ from those for tests not allowing calculator use. On completion of this portion of the project, it was noted that specifications written for calculator-based tests differed from those previously written in only two ways: (1) the complexity of the problems, and (2) the types of distractors. Because the problems reflected real-world situations containing realistic numbers, the increase in complexity was not surprising. The variation in the types of distractors was also not surprising given that one of the requirements originally placed on the calculator-based items was that they would contain distractors that reflect common calculator errors. Students without calculators would not be as likely to make these particular errors!

What was surprising, however, was the manner in which the calculator-specific errors were derived. It seems that students are likely to make two kinds of errors when using the calculator. One of these errors is mathematical; the other is mechanical. The mathematical error reflects illogical thinking or a misunderstanding of an algorithm. This type of error would occur much the same way it would occur without a calculator. The calculator merely makes it easier to get an answer. Consider the following problem:

> Last year the zoo sold 187 405 pounds of peanuts to visitors to feed the elephants. This year 224 721 pounds of peanuts were sold. How many more pounds of peanuts were sold this year?

A common mathematical error made by students is to apply the subtraction algorithm incorrectly by subtracting the smaller digit from the larger in each column regardless of its placement in the problem. Without a calculator, such an answer could be 163 324 pounds. With a calculator, these students would be just as likely to place the first number into the calculator and subtract the second one from it, thereby achieving an answer of −37 316 pounds and choosing that distractor. Both of these errors reflect a poor understanding of the subtraction process. Because of their previous training and experience with mathematical error patterns made by students, the writers used in this project had little difficulty identifying reasonable distractors in this category.

Mechanical errors are those made uniquely with the calculator—inadvertently pressing the wrong key, for example. Because they are not mathematical in nature, these errors seem to be more difficult for educators to predict.

After all item specifications were written, they were reviewed by a panel of teachers, supervisors, professors, and mathematics consultants and revised as necessary. Four test items were then written for each of the twenty-five specifications. Charts, graphs, and tables were inserted where the writers deemed most appropriate. Necessary changes to item specifications were made at the time of item writing to ensure that the test items would be as discriminating as possible. The 100 items were reviewed by the review panel, and revisions were made as necessary. The final set of test items was then prepared for use in the pilot-testing portion of the project.

The Pilot Test

Permission was obtained from a public school system in Florida to pilot-test items at grade levels 5, 8, and 11. Principals were asked to identify five students at each grade level who varied in their achievement in mathematics and in their experience with calculators. At the request of the individual schools, testing time was limited to one hour in the elementary schools and fifty minutes in the secondary schools.

Results of the Pilot

The pilot study was completed in an attempt to answer at least four questions:

1. *To what extent must previously used test items and specifications be altered in order to be appropriate for use on calculator-based tests?* Test items created for this project differed from those found on previous assessment tests used in Florida in two ways: (*a*) the numbers used were more appropriate to the real-world situations illustrated in the problems, and (*b*) distractors representing common calculator errors were present. Results of the pilot testing seem to indicate that both changes were appropriate.

2. *How long does it take to complete test items that are calculator-based?* Of the fifteen students tested, only two failed to complete the test in the alloted time. Because comparable tests were not administered to these students without the use of a calculator, results relative to this question are inconclusive. However, the most common comment made by eighth- and eleventh-grade students during the posttest interview was that the calculator actually slowed them down. In grade 11, two of the five students indicated that using the calculator made the test "quicker, not easier."

3. *To what extent does prior knowledge of calculators affect results?* When the calculators were distributed to the students at the various grade levels, it was noted that as the students' ages increased, so did their experience with calculators. The fifth-grade students were literally in awe while the calculator checks were being conducted, and after testing was completed, they indicated that they had never used a calculator in school. Results of testing indicated that the fifth-grade students made the majority of errors and that students in that group were the only ones who chose distractors that indicated calculator-based errors. The eighth graders used the calculator comfortably, and the eleventh-grade students began testing before the calculator checks were completed! It can be concluded from these data that

students perform in direct relationship to their experience with calculators on calculator-based tests.

4. *How does the availability of calculators during testing affect students' attitudes toward the testing situation?* Permission to use calculators while testing seems to improve students' attitudes toward the test situation. When students were interviewed about their attitude toward testing with and without calculators, only one eleventh-grade student, who omitted fourteen items and made twenty-three errors, indicated that the use of a calculator made her attitude more negative. The remaining fourteen agreed that the presence of the calculator made them feel more confident, and therefore more positive, about the testing situation.

CONCLUSIONS

Clearly, the calculator is a critically important tool with which students at all grade levels should be conversant and comfortable. As calculators become more successfully integrated into the mathematics curriculum, so too should they become a more valuable part of any program designed to assess the curriculum. The Florida experience indicates that it is possible to design appropriate items for calculator-based tests in mathematics at a variety of grade levels.

REFERENCES

Alberta Department of Education. *Guidelines for the Use of Calculators, Grades 1–12*. Edmonton, Alberta: The Department, 1981.

California State Department of Education. *Mathematics Frameworks for California Public Schools*. Sacramento: The Department, 1986.

Carpenter, Thomas P., Mary Kay Corbitt, Henry S. Kepner, Jr., Mary Montgomery Lindquist, and Robert E. Reys. "Calculators in Testing Situations: Results and Implications from National Assessment." *Arithmetic Teacher* 28 (January 1981): 34–37.

Carter, Betsy Y., and Steven J. Leinwand. "Calculators and Connecticut's Eighth-Grade Mastery Test." *Arithmetic Teacher* 34 (February 1987): 55–56.

Dye, David L. "The Use and Nonuse of Calculators on Assessment Testing." St. Paul, Minn.: Minnesota State Department of Education, 1981.

Florida Chamber of Commerce, Florida Education & Industry Coalition, and Florida State Department of Education. *A Comprehensive Plan: Improving Mathematics, Science, and Computer Education in Florida*. Tallahassee: Author, 1989.

Florida State Department of Education. *Student Performance Standards of Excellence for Florida Schools in Mathematics, Science, Social Studies, and Writing*. Tallahassee: The Department, 1984.

———. *Minimum Student Performance Standards of Florida*. Tallahassee: The Department, 1985.

Heid, M. Kathleen. "Calculators on Tests—One Giant Step for Mathematics Education." *Mathematics Teacher* 81 (1988): 710–13.

Kansky, Bob. "Technology Policy Survey: A Study of State Policies Supporting the Use of Calculators and Computers in the Study of Precollege Mathematics." Reston, Va.: National Council of Teachers of Mathematics, 1987.

Kenelly, John W., ed. *The Use of Calculators in the Standardized Testing of Mathematics*. New York: College Entrance Examination Board, 1989.

Kouba, Vicky L., and Jane O. Swafford. "Calculators." In *Results from the Fourth Mathematics Assessment of the National Assessment of Educational Progress*. Reston, Va.: National Council of Teachers of Mathematics, 1989.

Leitzel, Joan R., and Bert K. Waits. "The Effects of Calculator Use on Course Tests and on Statewide Mathematics Placement Tests." In *The Use of Calculators in the Standardized Testing of Mathematics,* edited by John W. Kenelly. New York: College Entrance Examination Board, 1989.

Long, Vena M., Barbara Reys, and Steven J. Osterlind. "Using Calculators on Achievement Tests." *Mathematics Teacher* 82 (1989): 318–25.

Michigan Department of Education. *Michigan Essential Goals and Objectives for Mathematics Education*. Lansing, Mich.: The Department, 1988.

National Council of Supervisors of Mathematics. "Essential Mathematics for the Twenty-first Century: The Position of the National Council of Supervisors of Mathematics." *Arithmetic Teacher* 37 (September 1989): 44–46.

National Council of Teachers of Mathematics. "A Position Statement on Calculators in the Mathematics Classroom." *Arithmetic Teacher* 34 (February 1987): 61.

————. "New Achievement Test Requires a Calculator." *NCTM News Bulletin* 27 (March 1991): 7.

North Carolina State Department of Public Instruction. *Mathematics Curriculum Study. A Report of the Mathematics Curriculum Study Committee to the North Carolina State Board of Education*. Raleigh, N.C.: The Department, 1983.

Oregon State Department of Education. *Mathematics. Grade 8. Oregon Statewide Assessment, 1985*. Salem, Oreg.: The Department, 1985.

Reys, Robert E. "Calculators in the Elementary Classroom: How Can We Go Wrong!" *Arithmetic Teacher* 28 (1980): 38–40.

Wilson, James W., and Jeremy Kilpatrick. "Theoretical Issues in the Development of Calculator-based Mathematics Tests." In *The Use of Calculators in the Standardized Testing of Mathematics,* edited by John W. Kenelly. New York: College Entrance Examination Board, 1989.

18

Calculators in State Testing: A Case Study

Steven J. Leinwand

TESTING has been and is now, perhaps more than ever, an enabling variable for curricular reform. Overwhelmingly, teachers teach what is assessed. When tests change—in content and in kind—curricular and instructional shifts follow. It is not surprising, therefore, that few issues in mathematics education today are as important or as controversial as the use of calculators on high-stakes tests.

This case study traces the successful campaign in one state to incorporate calculators on just such a high-stakes statewide test of mathematical achievement. It is written in the hope that Connecticut's experiences can help policy makers and mathematics educators across the nation better navigate the political, educational, and emotional minefields that surround this issue.

BACKGROUND

In 1984, the Connecticut General Assembly, at the urging of a new commissioner of education, agreed to replace the state's ninth-grade Proficiency Test with new fourth-, sixth-, and eighth-grade Mastery Tests. The proposal for a new testing program responded to calls for *earlier* testing of a *broader* range of skills with *higher* standards. The ninth-grade mathematics test emphasized arithmetic computation and had begun to suffer from ceiling effects. As in many states, the ninth-grade test represented Connecticut's response to the minimum-competency movement sweeping state legislatures in the late 1970s and thus predated national calls for a focus on problem solving and the availability of relatively inexpensive calculators.

Working cooperatively, the state department of education's assessment and curriculum bureaus responded to the 1984 legislative mandate by proposing the development of new tests in reading, language arts, and mathematics that—

- represented a much broader range of content than previously;
- assessed skills and concepts that teachers agreed were important, developmentally appropriate, and reasonable to teach;
- provided teachers, administrators, and other policy makers, as well as the public, with clear, useful data with which to evaluate and monitor programs, instruction, and student achievement;
- would help drive curriculum and instruction in positive directions.

To meet these goals in mathematics, Connecticut instituted a system of criterion-referenced tests based on a set of objectives developed by the state Mathematics Advisory Committee and validated through a survey of nearly 1000 teachers.

FROM IDEA TO DECISION

Since a mandate was given to "go out and develop the best test you can," it was not surprising that the issue of using calculators on at least the eighth-grade test surfaced early and often. Fortunately, Connecticut's 1981 *Guide to Curriculum Development in Mathematics* (Connecticut State Board of Education 1981), on which the tests were largely based, noted that "the widespread availability of calculators cannot be ignored when developing a mathematics curriculum" (p. 41).

The legitimate concerns that Connecticut teachers needed and deserved fair and adequate notice before allowing calculators to be used on the test were mollified by reference to the *Guide*. Available and widely disseminated for four years at the time, the *Guide* stated that "the inexpensive hand-held calculator dramatically alters the need for mastery of complex computation that often occupies months and even years of teaching. One of the strongest arguments for calculator use is the oportunity to increase significantly student exposure to real-world problems and applications" (pp. 41–42). Since this was exactly what the Mathematics Advisory Committee was trying to accomplish with this new test, the *Guide* proved to be an invaluable part of the foundation on which the final decision was based.

Just as the *Curriculum and Evaluation Standards for School Mathematics* (NCTM 1989) can be used today to support policy decisions of this sort, in 1984 we relied on NCTM's *Agenda for Action* (NCTM 1980). Recommendation 3, that "mathematics programs take full advantage of the power of calculators and computers at all grade levels," and Recommendation 5, that "the success of mathematics programs and student learning be evaluated by a wider range of measures than conventional testing," were indispensible in helping win official support for what became known as the "calculator initiative." In an environment where people want to be current and truly want to serve the best interests of students, NCTM's proclamations, agendas, and standards can play a crucially supportive role. So the necessary groundwork had been laid.

Among the mathematics and assessment professionals involved in developing the Mastery Tests, the inclusion of calculators seemed like a good idea whose time had come. But decisions of this sort are ultimately political, and not educational, matters. Accordingly, we faced the dual needs to (1) justify more clearly and publicly the use of calculators on the test and (2) gather and present more compelling data on the issue before seeking support and final approval from the commissioner of education and the Connecticut State Board of Education. In hindsight, our most critical activities were the preparation of a position paper and a survey of teacher attitudes.

With advice from the Mathematics Advisory Committee, the mathematics consultants from the department of education drafted a position paper on "Calculators and the Eighth-Grade Mastery Test" (see Carter and Leinwand [1987]). The paper began with a recommendation that hand-held calculators be made available to all eighth-grade students for use on the problem-solving/applications and measurement/geometry items of the forthcoming eighth-grade Mastery Test. The primary rationale for this recommendation was that the state's testing program must respond to the increasing use of the calculator as a computational tool in our society. It was also argued that the program must send a clear message to the educational community and to the public that the emphasis of mathematical instruction, particularly at the upper elementary and middle school levels, must shift toward critical thinking skills and problem solving and away from a predominantly computational focus.

The position statement concluded by summarizing that allowing students to use calculators would result in—

- a strong, sound, and consistent statement about what is important in mathematics instruction;
- a stronger, more realistic, and responsive testing program;
- the equitable provision of calculators for use by each eighth grader in the state.

This position paper and the theme of state-level leadership on which it was built helped to convince division directors, deputy commissioners, and finally the commissioner of the value of this initiative.

As part of building a statewide consensus on the content of the tests, the state Mathematics Advisory Committee surveyed teacher attitudes on the specific objectives planned for the test and on the issue of calculator use. More than 700 sixth-, seventh-, and eighth-grade teachers responded to the survey questionnaire, and we disaggregated the results to study the differences, if any, between teachers in the state as a whole and teachers in

Connecticut's five largest urban districts. Survey results on the four calculator questions are shown in figure 18.1.

Teacher Survey of Preliminary Objectives
Connecticut Grade 8 Mastery Test

On the pilot test, students will have access to calculators for the problem solving/applications and measurement/geometry items. Please respond to the following questions regarding calculator usage in school and on tests.

	% all respondents (n = 724)	% respondents in 5 largest cities
1. Have your students used calculators in their classes?		
Frequently	32.0	52.8
Occasionally	62.3	45.8
Never	5.7	1.4
2. Do you or would you encourage students to use calculators in school?		
Yes, in all situations	9.7	15.1
Yes, in certain situations	87.6	84.9
No, never	2.7	0.0
3. Do you or would you encourage students to use calculators when doing their homework?		
Yes, in all situations	13.4	23.3
Yes in certain situations	82.7	75.3
No, never	3.9	1.4
4. How do you feel about allowing students to use calculators on the problem solving/applications and measurement/geometry items on the Eighth-Grade Mastery Test?		
I think it's a good idea	18.9	23.0
I'm not sure	35.5	32.4
I'm definitely against it	45.6	44.6

Fig. 18.1

As expected, we found that calculators *were* being used in nearly all classrooms and that large majorities of teachers encouraged their students to use calculators in school and on homework. However, when teachers were asked directly how they felt about allowing students to use calculators on the new test, fewer than one in five thought it was a good idea, nearly two in five weren't sure, and almost half were definitely against it! As it turned out, one of the most powerful aspects of the survey was the fact that urban teachers were consistently *more* likely to use and encourage calculators.

Armed with the commissioner's backing, the position paper, and the survey results, we turned to the last hurdle—the Connecticut State Board of Education. When we consider the political realities of such policy-making boards, it should not have been surprising that the decision to proceed was not ultimately made on the basis of educational grounds. In fact, a number of board members expressed lingering concerns about the wisdom and the appropriateness of letting students use calculators at all, let alone on a state test. However, after lengthy discussion, the unanimous vote appeared to come down to four key factors:

1. *Reasonableness*. The plausibility of the arguments in the position paper and the lack of any real public or editorial outcry helped to shift the initial sense that this was a radical change to an open-minded willingness to seriously consider the issue. After reflection and discussion, state board members and others realized that this was not as outrageous a request as it might at first have sounded.

2. *Balance*. The board was also clearly comforted that calculators were being introduced in a balanced fashion. That they would be available for a little less than half the test items and that traditional computational skills were still being tested without calculators convinced people that we were not throwing out the baby with the bathwater.

3. *Leadership*. Fortunately, the Connecticut State Board of Education sees itself as a national leader in the field of education. The board has been extraordinarily supportive of progressive initiatives and was therefore positively disposed to being once again a national trend setter.

4. *Equity*. Perhaps the most important factor underlying the decision was the issue of equity. Several board members expressed deep concern that calculators would be used and encouraged increasingly in Connecticut's wealthier and suburban districts, whereas far fewer urban students would have these opportunities. But the board appeared to reason that prohibiting calculators on the state test would exacerbate these differences in opportunity. On the contrary, permitting and even purchasing calculators would be seen as a powerful statement on behalf of providing equal educational opportunity and narrowing the gap between rich and poor.

With the state board of education's vote in December 1985 to allow access to calculators on part of the eighth-grade Mastery Test and to purchase 35 000 calculators (one for every eighth grader), Connecticut became the first state to sanction the use of calculators on a statewide, high-stakes, legislatively mandated census test of student achievement.

FROM DECISION TO IMPLEMENTATION

Parallel with the activities described above, state department staff and the state Mathematics Advisory Committee were moving ahead with the development of the entire test. This process produced a list of thirty-six objectives representing the mathematics that we believe beginning eighth graders should have mastered (see fig. 18.2). The committee recommended, and more than 700 teachers validated, eleven objectives in the conceptual understandings domain, ten in the computational skills domain, ten in prob-

Eighth-Grade Mastery Test Objectives

Conceptual Understandings*

1. Order fractions
2. Order decimals
3. Round whole numbers
4. Round decimals to the nearest whole number, tenth, and hundredth
5. Multiply and divide whole numbers and decimals by 10, 100, and 1000
6. Identify fractions, decimals, and percents from pictorial representations
7. Convert fractions to decimals and vice versa
8. Convert fractions and decimals to percents and vice versa
9. Identify points on number lines, scales, and grids
10. Identify ratios and fractional parts from given data
11. Identify an appropriate procedure for making estimates with decimals and fractions

Computational Skills*

12. Add and subtract whole numbers less than 10 000
13. Multiply and divide two- and three-digit whole numbers by one- and two-digit numbers
14. Add and subtract decimals (to hundredths) in horizontal form
15. Identify the correct placement of the decimal point in multiplication and division of decimals
16. Add and subtract fractions and mixed numbers
17. Multiply fractions and mixed numbers
18. Determine the percent of a number
19. Estimate sums and differences of whole numbers and decimals including making change
20. Estimate products and quotients of whole numbers and decimals
21. Estimate fractional parts and percents of whole numbers and money amounts

Problem Solving/Applications (with calculator available)

22. Compute sums, differences, products, and quotients using a calculator
23. Interpret graphs, tables, and charts
24. Solve one- and two-step problems involving whole numbers and decimals including averaging
25. Solve one- and two-step problems involving fractions
26. Solve problems involving measurement
27. Solve problems involving elementary probability
28. Estimate a reasonable answer to a given problem*
29. Solve problems with extraneous information
30. Identify needed information in problem situations
31. Solve process problems involving the organization of data

Measurement/Geometry (with calculator available)

32. Identify figures using geometric terms
33. Measure and determine perimeters and areas
34. Estimate lengths, areas, volumes, and angle measures
35. Select appropriate metric or customary units and measures
36. Make measurement conversions within systems

*Will be tested without calculators available

Fig. 18.2

lem solving/applications, and five in measurement/geometry, with calculators available for the latter two domains.

Next came item development. The criterion-referenced design of the test called for four items for each objective. Here we found ourselves in a real bind that others, with better planning, should be able to avoid. Our problem was that the decision to use and purchase calculators was not yet made, and we were under time pressure to write and pilot the items. Given these circumstances, we walked a tightrope and developed items where a calculator would be helpful but not indispensible. This was only a minor concern, since the primary purpose of allowing calculators in the first place was to increase the likelihood of their use in ongoing classroom instruction and remove the excuse that "the state doesn't allow them, so why should I?"

The closest we came to calculator-active items was in the objective involving the computation of sums, differences, products, and quotients using a calculator. This objective was included to determine whether students had basic knowledge of how to use a calculator. The test also included several word problems where large numbers made the problems more realistic and the calculator more helpful. Figure 18.3 displays several of these items.

Although a "go" decision was never a foregone conclusion, it often seemed during the decision-making process that *supplying* the calculators and *training* students were larger concerns than simply allowing the *use* of calculators. To address these concerns, we (1) carefully constructed calculator familiarity exercises as part of the test administration, (2) designed a student questionnaire about experience with calculators as part of the pilot testing, and (3) engaged in a formal vendor-selection process for providing the calculators.

In September 1986, more than 30 000 Connecticut eighth graders were provided with state-purchased calculators for use on the problem solving/applications and the measurement/geometry items of the new Grade 8 Connecticut Mastery Test.

OUTCOMES

Perhaps the biggest surprise of the entire calculator initiative was how quickly the entire effort became "one big ho-hum." Although the decision process never induced a great deal of controversy, it was not popular in all circles. Several editorials and editorial cartoons, some teachers, and a few administrators had publicly expressed varied concerns. However, within weeks of the initial problem-free administration, the entire issue evaporated, and the change was painlessly institutionalized. Even when the replacement by the state of lost or broken calculators was discontinued in 1988 and districts became responsible for providing all the calculators, hardly anyone raised concerns or questions about what we were doing.

$662.4 \div 3.2 =$

f 207 g 20.7 h 288 j 659.2

$235.56 - 73.2 + 1912 - 786.35 =$

f 2860.71

g −432.79

h 1288.01

j 1434.41

The record shop sold 137 albums on Wednesday, 15 albums on Thursday, 214 albums on Friday, and 31 albums on Saturday. What was the average number of albums sold each day?

f 205 g 820 h 82 j 3280

The softball team brought 18 new uniforms at $23.75 each. If their budget was $650, how much money did they have left?

f $427.50

g $222.50

h $1077.50

j $293.75

Use the information below to answer the question.

Wheeler's Tool Company
Yearly Profit & Loss Statement

	1983	1984
1. NET SALES	$492,120	$689,270
2. Cost of Goods Sold	$358,315	$599,665
3. GROSS PROFIT	$133,805	$89,605
4. Other Operating Expenses	$57,536	$43,906
5. PROFIT BEFORE TAXES	$76,269	$45,699
6. Taxes	$25,169	$15,081
7. NET PROFIT	$51,100	$30,618

How much greater was the company's GROSS PROFIT in 1983 than in 1984?

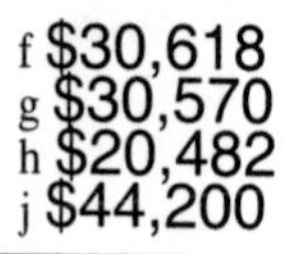

f $30,618

g $30,570

h $20,482

j $44,200

Fig. 18.3

Overall, we have found that people's fears were unfounded. Computation scores have not declined in the years the test has been administered, but our hopes for significantly higher problem-solving scores have also not materialized. We have anecdotal data suggesting that many more seventh- and eighth-grade teachers are using calculators in their classes, and we hear that there is a concomitant reduction in emphasis on paper-and-pencil computation in these grades.

As for hard data, we have demonstrated conclusively that nearly all students can use a calculator to find sums, differences, products, and quotients. This objective, entailing three-, four-, and five-digit numbers, has consistently been mastered (three or four out of the four items) by 98 percent of Connecticut eighth graders. For example, 98 percent of the students tested can correctly add numbers like 952.4 + 24.1 + 5.2 + 501.8. Data like these allowed us to delete the fifteen-minute calculator-usage instruction part of the test administration after the first year of testing. In addition to saving fifteen minutes, teachers through their comments and students through the test results told us that such instruction simply was not needed.

We have also learned that when Connecticut eighth graders incorrectly answer a one- or two-step word problem involving two- or three-digit numbers, the problem is almost certainly answered incorrectly because of a failure to choose the correct operation, not because of computational errors. On other tests, this information is often more difficult to ascertain.

For example, consider this item: "The library purchased 9 computer program disks at $38.97 each. How much did the library have left from a budget of $500?" Even with calculators, the results are both telling and disappointing. In 1989, 74 percent of the eighth graders got this item correct, but 10 percent simply multiplied 9 × 38.97 and 15 percent simply subtracted 38.97 from 500. We have used results like these to encourage more forcefully the use of calculators as vehicles for reducing time and emphasis on decimal and whole number computation and increasing time, opportunity, and emphasis on applications and one- and two-step word problems.

CONCLUSION

Connecticut's successful campaign to include calculators on the state test and the speed with which this policy has been accepted and institutionalized can provide other states and districts with some rough guidelines for pursuing similar policies:

- Build the case for calculators cogently and with broad support, using position papers, white papers, survey data, and so on.
- Focus as much on emotional and political issues like equity as on educational or mathematical issues.

- Be willing to compromise and take "half a loaf" the first time around in order to get your foot in the door.
- Seriously consider purchasing or underwriting the purchase of the initial batch of calculators.

Today Connecticut is planning the next generation of Mastery Tests, including expanding calculator access both within the eighth-grade test and into the sixth-grade test. We expect to reduce isolated computation further and embed as much of the mathematics as possible in realistic contexts. In other words, we believe that the inclusion of calculators on our current test has served as a key enabling variable for teachers and districts now striving to implement the vision represented by the *Curriculum and Evaluation Standards for School Mathematics* (NCTM 1989). Rather than face the obstacles of calculator prohibitions on high-stakes tests, Connecticut's middle and junior high school teachers for the last five years have been encouraged to prepare students for calculator-active assessment.

REFERENCES

Carter, Betsy Y., and Steven J. Leinwand. "Calculators and Connecticut's Eighth-Grade Mastery Test." *Arithmetic Teacher* 34 (February 1987): 55–56.

Connecticut State Board of Education. *A Guide to Curriculum Development in Mathematics.* Hartford: The Board, 1981.

National Council of Teachers of Mathematics. *An Agenda for Action: Recommendations for School Mathematics of the 1980s.* Reston, Va.: The Council, 1980.

———. *Curriculum and Evaluation Standards for School Mathematics.* Reston, Va.: The Council, 1989.

19

The Effects of Calculators on State Objectives and Tests

Joseph N. Payne

AT ABOUT the same time that the National Council of Teachers of Mathematics was developing the *Curriculum and Evaluation Standards for School Mathematics* (1989), Michigan was engaged in a major rethinking of its mathematics curriculum. The existing objectives, first written in the early 1970s, were minimal. Much discussion had taken place throughout the state about the need for a more comprehensive set of objectives that would reflect the mathematical needs of a consumer-oriented and technological society.

Early in 1986, the Michigan Education Assessment Program (MEAP) gave a small grant to the Michigan Council of Teachers of Mathematics (MCTM) to prepare new mathematics objectives as a model for instructional guidelines. It was understood that the state assessment done yearly at the beginning of grades 4, 7, and 10 would be revised in accordance with this new instructional framework. The draft of the objectives was prepared by a committee of twelve mathematics education specialists and various subcommittees that included teachers. The draft was then reviewed widely in the state, revised by the committee, and approved by the Michigan State Board of Education in March 1988 (Michigan Department of Education 1988). An extensive interpretive document written by a similar group from MCTM and published in March 1989 (Michigan Department of Education 1989) contains rationale for the objectives and the content strands, explanations of the objectives, sample problems, and suggestions for specifications for grade levels, K–3, 4–6, and 7–9. The objectives are intended to set new directions for mathematics education in Michigan.

IMPORTANCE OF CALCULATOR USE

It was the firm view of the committee that major changes in the mathematics curriculum could come about only with a reduction in expectations for paper-and-pencil computation. Consequently, MCTM insisted, as a condition in its agreement to prepare the objectives, that calculator use be assumed as integral parts of instruction and testing. The committee knew that K–8 textbooks contain a preponderance of paper-and-pencil computation exercises, reflecting current classroom practice. This conclusion is corroborated in a study by Porter (1988) in which he found that from 70 percent to 75 percent of instructional time in grades 3, 4, and 5 was devoted to computational skills. In one class, it was more than 90 percent.

THE IMPACT OF CALCULATOR USE ON OBJECTIVES

Once calculators are an integral part of instruction and testing, the amount and type of computation is modified. In the new Michigan objectives, there is substantial reduction in the amount of paper-and-pencil computation, yet needed computational skills are preserved. Mental arithmetic and estimation are major strands. Calculator literacy is a strand, and computation to be done with calculators is defined explicitly.

Beyond the computation modifications, other major goals have been accomplished. Conceptual knowledge receives much greater emphasis, including the meaning of numbers, computational algorithms, and other mathematical topics. Problem solving and applications are included in all content areas. There is a substantial increase in the amount of content for the oft-neglected areas of geometry, graphical representation, statistics, and algebraic ideas.

The objectives are organized by content strands. Each content objective is identified by a mathematical process to describe the real intent of the objective and, in effect, the way it can be assessed. Because the process was identified at the time the objective was written, descriptions of each mathematical topic are thorough and specific. The result is a well-rounded view of each mathematical topic.

The content-process matrix shown in figure 19.1 is called the Michigan Framework. The figure also shows which clusters of objectives are assessed with or without calculators available.

Computation Expectations

The objectives labeled Computation are paper-and-pencil objectives. This relatively narrow view of computation, in contrast to a broader view that includes mental arithmetic and estimation, was done deliberately so that there would be clear expectations to reduce the dominance of paper-and-pencil computation in school mathematics. The separation of computation from calculator and computer use (very little was done with computers because there is a separate set of Michigan computer objectives) was intended as an added impetus for changes in the mathematics curriculum by making explicit the computation that is to be done by calculators.

Mathematical Content Strands	Conceptualization	Mental Arithmetic	Estimation	Computation	Applications and Problem Solving	Calculators and Computers (use)
Whole Numbers and Numeration	□	■	■	■	□	□
Fractions, Decimals, Ratio and Percent	□	■	■	■	□	□
Measurement	□	■	□	□	□	□
Geometry	□	■	■	□	□	□
Statistics and Probabitity	□	■	□	□	□	□
Algebraic Ideas	□	■	■	□	□	□
Problem Solving and Logical Reasoning	□	□	□	□	□	□
Calculators (literacy)	■	■	■	■	■	■

Legend □ Calculator Use Permitted
■ Calculator Use NOT Permitted

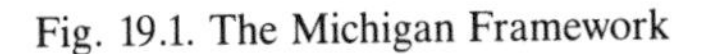

Fig. 19.1. The Michigan Framework

The inclusion of the calculator as a parameter for instructional objectives and for testing has its greatest impact on the objectives for numbers—whole numbers, fractions, decimals, ratio, and percent. Major changes in objectives related to number include some limitations on the size of numbers for paper-and-pencil computation and the development of strong strands for mental arithmetic and estimation. Expectations for whole number computation are limited to adding four three-digit numbers, subtracting two three-digit numbers, multiplying two-digit and three-digit numbers, and dividing by numbers 30 or less or multiples of 10 to 100. There are no whole number paper-and-pencil computation objectives for grades 7–9, a major change in the usual content in textbooks and taught at these levels.

Computation objectives for decimals have the same limitations on digits as whole numbers. The complexity of fraction computation is reduced substantially, and some work is shifted to later grade levels. There is extensive use of easily recognized fractions and related fractions such as halves-fourths-eighths, thirds-sixths, and fifths-tenths. Objectives for ratio and percent are for the most part conceptual, with only one computational objective for ratio. There are no paper-and-pencil computation objectives for percent, reflecting a monumental change in the way percent content is to be taught in grades 6–9.

Mental Arithmetic and Estimation

There are mental arithmetic objectives for all classes of numbers. Mental arithmetic includes adding and subtracting two-digit numbers, multiples of 10 to 100, and multiples of 100; multiplying multiples of 10 and 100 and two-digit numbers, and some special products; dividing multiples of 10, 100, and 1000; equivalent fractions and operations on fractions and decimals such as $\frac{1}{2} + \frac{1}{4}$ and $1 - 0.3$; fraction-decimal-percent equivalents for easily recognized fractions; and finding 1%, 10%, 25%, and 50% of a number.

The intent of estimation objectives is to see that answers are reasonable and measures are sensible. Estimates for whole numbers include rounding and operations on larger numbers. For fractions, decimals, and percent, estimation includes the size of regions, equivalence, powers of 10, and operations that are more complex.

The objectives for numbers show clearly the intent that calculators should be an integral part of instruction. Much more attention is given to the conceptual understanding of numbers and operations on numbers through the use of models and other well-understood constructs. Computation is limited. Mental arithmetic and estimation are prominent.

Calculator Literacy

The calculator literacy strand contains three major objectives related to calculator keys, computation, and the limitations of calculators.

The first major objective is *to recognize specific calculator keys.*

The second major objective is, in fact, a summary of the calculator-use objectives embedded in each of the content strands: *To perform appropriate computations with a calculator.*

The third major objective deals with limitations and calculator display: *To recognize certain common limitations to calculators and be able to interpret selected calculator-displayed symbols.*

THE MICHIGAN TEST

As a part of the overall plan to develop a state test based on the objectives, a committee of mathematics education specialists and classroom teachers reviewed the objectives and made recommendations on what percentage of the test was to be allocated to each content strand. These allocations are shown in figure 19.2.

	Grade 4	Grade 7	Grade 10
Whole numbers and numeration	38%	10%	3%
Fractions, decimals, ratio, and percent	5	20	25
Measurement	12	15	10
Geometry	13	15	14
Statistics and probability	7	10	14
Algebraic ideas	6	10	17
Problem solving and logical reasoning	15	15	15
Calculator literacy	4	5	2

Fig. 19.2. Percentage allocations of content objectives for the state test

The graph in figure 19.3 shows the resulting distribution by mathematical process and by grade levels (Michigan Educational Assessment Program 1990, p. 10). The graph reflects the relative importance of each mathematical process and changes in emphasis from level to level. Conceptualization comprises about half the objectives in grades K–4, about a third of them in grades 4–6, and about a fourth in grades 7–9. Mental Arithmetic and Estimation combined increases from about 15 percent of objectives in grades K–3 to 25 percent in 4–6 and 30 percent in 7–9, reflecting the broader range of content for mental arithmetic and estimation available at succeeding grade levels. Paper-and-pencil computation is about 15 percent at K–3 and 4–6 and about 10 percent at 7–9. Applications and Problem Solving show increases from about 15 percent to 25 percent. Calculator literacy objectives increase but occupy only a small percentage of the overall objectives.

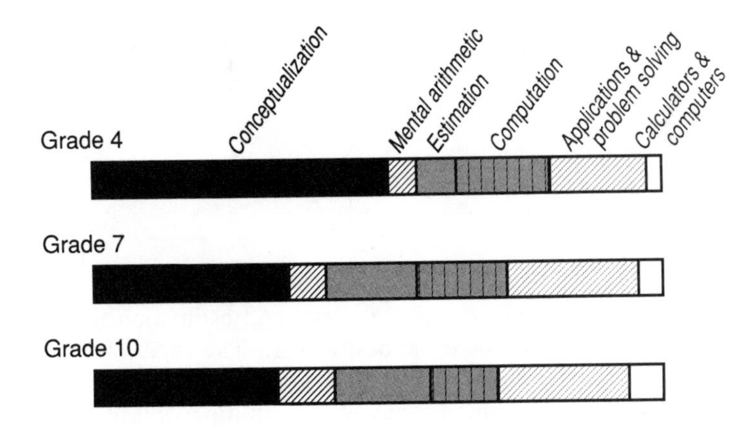

Fig. 19.3. Distribution of objectives by mathematical processes for the state test

The test was first administered in the fall of 1991. The test included a section on mental arithmetic and estimation, a section on paper-and-pencil computation with no calculators allowed, and a section with calculators permitted including one open-ended problem. In general, the test is a power test with no time limits, so that students can work at their own pace. The entire test is expected to be given in about two hours. The only timed portion of the test is the section on Mental Arithmetic and Estimation.

Mental Arithmetic and Estimation Section

The first section of the test is a timed, multiple-choice test on mental arithmetic and estimation with no calculators allowed. Students are told not to work the problems on scratch paper or in the test booklet and to use their pencils only to mark the answer sheet. The space surrounding each problem has a gray screen to discourage written computation. In the pilot tests, students and teachers were very conscientious about adhering to the directions for doing the work mentally. Sample items appropriate for grade 7 are shown in figure 19.4.

Lisa and her friends ordered 2 pizzas. Lisa ate one fourth of a pizza and her friends ate the rest. How much pizza did her friends eat?

○ $\frac{1}{2}$

○ $1\frac{3}{4}$

○ $1\frac{1}{4}$

○ $2\frac{3}{4}$

12.2×3.09 is between

○ 0 and 25

○ 25 and 50

○ 50 and 100

○ 100 and 500

$\frac{3}{8} + \frac{1}{4} = ?$

○ $\frac{4}{12}$

○ $\frac{4}{8}$

○ $\frac{5}{4}$

○ $\frac{5}{8}$

Fig. 19.4. Sample Mental Arithmetic and Estimation items for grade 7

In earlier tryouts of the mental arithmetic and estimation test, various formats were used to determine the best format. The experimental results from several different classrooms indicated that student performance under timed administration of printed items with a multiple-choice format was indistinguishable from untimed, oral presentations. The multiple-choice format was much better for estimation items because in the open-ended format students had great difficulty in knowing how close their estimates should be.

No Calculators Allowed Section

The second section of the test, untimed, deals with computation to be done with paper and pencil with no calculators allowed. It is a relatively short section because of the small number of paper-and-pencil computation objectives. The format is multiple choice. Students are told that they can use the space in the test booklet for scratch paper. In the test pilot, almost all students finished in fifteen minutes or less.

Calculators Allowed Section

The third, untimed, section of the test is by far the longest, with items for all the other objectives, including all problem-solving and conceptual objectives. Students are told that they will use paper, pencil, and other tools (ruler, protractor, clear-plastic grid) to solve the different kinds of problems. They are told that they are permitted to use a calculator to help answer any of the questions in this section. A few questions in this section ask students to find an answer by using a calculator, but many items are calculator independent. By observing students taking the test, teachers can see if students are making intelligent decisions on when to use or not use a calculator.

PRELIMINARY TEST RESULTS

When the objectives were written, it was assumed that each classroom would have a set of calculators, all alike, to facilitate instruction and testing. Because of cost and logistics, it was not possible for the state to provide calculators for the state test. Consequently, each school had to decide how calculators would be furnished.

Calculator Availability

In a survey of approximately 100 classrooms at each level in a pilot test, the teachers were asked about the provision of calculators by the school district for instruction and for testing. The teachers reported that about half the school districts furnish calculators for instruction for grades 4 and 10 and two-thirds of the districts do it for grade 7. In a similar question about the school district providing calculators for the test, about three-fourths of the districts provided them for grades 4 and 7 and slightly more than half for grade 10. For classes where districts did not provide calculators for the test, students brought their own. In one grade 4 class that was observed, all students brought their own calculator, resulting in more than a dozen different types of calculators used during the test.

Mathematical Skills and Understanding

A complete analysis of strengths and weaknesses by content strands and by mathematical processes will be done for the test data. While awaiting those results, some pilot-test data are revealing. It came as a surprise that results were so poor on certain computation items *with calculator available.* For example, fewer than 50 percent of tenth-grade students correctly found 1.18% of 3.4 or $4.56 \times \frac{18}{25}$, and barely 50 percent could express the fraction $\frac{51}{68}$ as a percent—all items on the part of the test where calculators are permitted. It seems obvious that a major educational initiative is needed for calculator literacy. Students need direct instruction on keys and keystroking. Calculators need to be used on a regular basis and permeate instruction in a substantial sense, which apparently has not happened on a large scale.

An examination of results from items on other content strands and mathematical processes points to other tentative conclusions. Estimation, an essential tool to check for keystroking or registry errors on calculators, needs major attention. With poor performance on simple questions about fractions, decimals, and percent, there appears to be a major need to strengthen the conceptual underpinnings for these numbers. There is a strong indication that changes are essential in the relative amount of instructional time allocated to the various mathematical processes, with less instructional time given to paper-and-pencil computation.

The results from the 1991 test with all students in grades 4, 7, and 10 will provide firmer conclusions as well as a basis for identifying strengths and weaknesses of the mathematics curriculum. It will be interesting to see results that are related to content that is newer to the curriculum—transformational geometry, statistical ideas such as box and stem-and-leaf plots, and algebraic ideas. Indications for needed curriculum changes and professional development for teachers will be clearer.

INSTRUCTIONAL SUPPORT

The Michigan Council of Teachers of Mathematics prepared a full day's workshop for teachers with accompanying written materials entitled "New Directions in Mathematics Education in Michigan" and directed by this author. MCTM trained 125 teachers and consultants in the state as leaders to conduct the workshops, and approximately fifty workshops with about thirty teachers each have been conducted. The workshop is intended to convey the content and the spirit of the Michigan objectives as well as acquaint teachers with national trends. Calculator use is one major feature of these workshops as well as the newer directions of concepts, manipulatives, mental arithmetic, and estimation and of strands such as geometry, statistics, and algebra. In each of the breakout sessions for grade K–3, 4–6, and 7–9 teachers, activities are used to help teachers see both the content of the objectives and the instructional directions. Teachers who have been to the workshops emerge with a view that major changes are coming in the mathematics curriculum and in instructional methods.

Another major initiative by MCTM through the Michigan Department of Education is entitled "Michigan Mathematics In-Service Project." Extensive materials have been prepared for elementary school teachers at two levels, grades K–2 and 3–6. The materials are designed to support a full-semester professional development program of approximately forty hours. The major purpose of the materials is to implement the Michigan mathematics objectives. The efforts of this newer project will become more evident during the 1991–92 school year. Preliminary results from the pilot test suggest that similar materials are needed for grades 7–9.

SUMMARY

The assumption of calculator use for instruction and for testing has indeed provided the opportunity to broaden the content objectives in mathematics and to reduce the emphasis on paper-and-pencil computation. The overall result is a set of objectives and statewide assessment using calculators that is more closely in tune with the personal and work-related needs of citizens in the twenty-first century. Whether these objectives and the resulting tests will have the intended influence on school mathematics in Michigan is yet to be determined. Results from the pilot test indicate that there is much to be done in redesigning the mathematics curriculum and the way mathematics is taught. The extensive activity under way now in the state suggests that results may be improved in the future.

REFERENCES

Michigan Department of Education. *An Interpretation of the Michigan Essential Goals and Objectives for Mathematics Education.* Lansing, Mich.: Michigan Council of Teachers of Mathematics, 1989.

———. *Michigan Goals and Objectives for Mathematics Education.* Lansing, Mich.: The Department, 1988.

Michigan Educational Assessment Program. *Essential Skills Mathematics Test Blueprint.* Version 2.1. Lansing, Mich.: Michigan Department of Education, 1990.

National Council of Teachers of Mathematics. *Curriculum and Evaluation Standards for School Mathematics.* Reston, Va.: The Council, 1989.

Porter, Andrew C. *A Curriculum Out of Balance: The Case of Elementary School Mathematics.* Research Series No. 191. East Lansing, Mich.: Michigan State University, Institute for Research on Teaching, 1988.

20

The Use of Calculators on College Board Standardized Tests

Carole E. Greenes
Gretchen W. Rigol

STANDARDIZED tests not only assess student abilities but also reflect accepted curricula and instructional methodology. For example, the recent emphasis on statistics in the mathematics curriculum (National Council of Teachers of Mathematics [NCTM] 1989) is mirrored in the increased number of items dealing with the interpretation of graphs on the College Board's Descriptive Tests of Mathematics Skills and the Mathematics Achievement Tests. The increasing use of calculators in mathematics instruction is being reflected in new College Board tests that require or permit the use of calculators.

Since the mid 1980s, there has been a groundswell of support for the use of calculators in both instruction and assessment (Mathematical Sciences Education Board 1990; NCTM 1989, 1986). This support has resulted from both the recognition of the usefulness of calculators in teaching mathematical concepts and in enhancing mathematical reasoning abilities and the increasing availability of calculators at low cost. In 1976, the College Board's Mathematical Sciences Advisory Committee (MSAC) discussed and explored ways to encourage the use of calculators for instruction and testing. Four years later, MSAC advocated the use of calculators on Advanced Placement Calculus Examinations. In 1983, the College Board, in its book *Academic Preparation for College,* listed the ability to use calculators as one of the basic academic competencies required of all students. In its *Academic Preparation in Mathematics* (College Entrance Examination Board 1985), the College Board gave additional testimony to the importance of knowing how and when to use calculators and how their use would affect the mathematics curriculum. Now, in the 1990s, the College Board is developing new tests at a variety of levels that require the use of calculators.

Although the use of calculators in teaching and testing has been endorsed by various professional organizations, including the National Council of Teachers of Mathematics (1986, 1989), the American Association for the

Advancement of Science (1989), and the Mathematical Sciences Education Board (1990), calculators are not universally used by American mathematics teachers. The College Board acknowledges this difference of opinion and is sensitive to educational concerns—particularly issues of equity—that are raised by the use of calculators in the teaching and testing of mathematics.

The equity issue concerning the accessibility of calculators was discussed in detail at the 1986 Symposium on the Use of Calculators in Standardized Testing, cosponsored by the College Board and the Mathematical Association of America. Participants concluded that "falling prices have essentially resolved the old equity issue of student access to calculators, and have introduced instead a more important issue of student access to adequate preparation for using these devices appropriately and well" (Kenelly 1989). In other words, the equity issue is a question of *equal access to proper instruction with calculators* rather than access to the devices themselves. Unfortunately, access to proper instruction is impaired by the paucity of major textbooks and programs in which the use of the calculator is central to facilitating the understanding of key mathematical concepts and processes. Without such instructional materials, learning how and when to use calculators is dependent solely on the inventiveness of the teacher and his or her policy about calculator use for classwork, homework, and tests. Students who have not been taught when and how to use calculators face an unfair disadvantage compared to calculator-sophisticated students.

In addition to instructional equity, equity concerns related to the level of complexity of calculators still exist. Should standardized tests be designed to reflect the use of "state of the art" calculators (e.g., graphing calculators), or should tests be based on calculators used by the majority of students (e.g., scientific calculators)?

THE USE OF CALCULATORS ON COLLEGE BOARD TESTS: 1976–1990

After seven years of research, the use of scientific calculators on Advanced Placement Calculus examinations was permitted (not required) on an experimental basis during the 1983 and the 1984 administrations of the examinations. Although the examinations were designed so that calculators were not needed, some students indicated that they had used their calculators on a substantial number of items. On the average, those students performed less well than students who used their calculators very little. Because equality and fairness (of experience) for all students could not be ensured, the policy of allowing calculators to be used on the AP Calculus examinations was suspended in 1984. The AP Calculus Committee also was concerned that the rapidly changing technology would result in great disparities among calculators in a very short time. The advanced technology no doubt would produce major changes in curricula, which would subsequently require substantive changes in test questions (Jones 1991). Despite these concerns, the various College Board mathematics committees continued to affirm their commitment to the use of calculators in classroom instruction and testing and to the development of standardized tests that permit or require the use of calculators.

On the basis of results of the experimental AP examinations, the Mathematical Sciences Advisory Committee (MSAC) of the College Board and the Achievement Committee recommended the development of an experimental Mathematics Achievement Test with items specifically designed to require the use of a calculator. In 1985, preparation began for the Mathematics Level IIC Achievement Test, which required the use of calculators. Since that time, several conferences, projects, and surveys of the College Board (e.g., the 1986 Symposium on the Use of Calculators in Standardized Testing; the 1990 New Possibilities Project; the 1990 AP Calculus Calculator Impact Study) have addressed technology issues and their impact on curriculum and assessment.

THE USE OF CALCULATORS ON COLLEGE BOARD TESTS: 1991–1994

With implementation dates ranging from the spring of 1991 to the fall of 1994, most mathematics tests offered by the College Board will either permit or require the use of calculators. A new version of the Mathematics Level II Achievement Test, the Level IIC, which requires the use of nonprogrammable scientific calculators, was first administered in June 1991 as an alternative to the noncalculator Level II Achievement Test. The Advanced Placement Calculus examinations (AB and BC) will require scientific calculators beginning in May of 1993. In the fall of 1993, students taking the PSAT/NMSQT (Preliminary Scholastic Aptitude Test/National Merit Scholarship Qualifying Test) will be permitted to use calculators. Calculators also will be allowed on the new Scholastic Aptitude Test (SAT I), which will be offered first in the spring of 1994.

Although the following sections discuss each major testing program separately, survey and experimental research results conducted by one program were taken into account by other programs in developing and determining their calculator policies. Differences in test content, purpose, population, and the method of administration contributed to the different implementation schedules and somewhat different calculator policies. With changing technology and the increasing use of calculators in instruction, calculator policies of the various College Board tests will be modified to reflect these changes.

Advanced Placement Calculus

In the spring of 1990, to study the impact of the use of calculators on the performance of students on AP Calculus tests, the College Board commissioned a special study (Morgan and Stevens 1991). Two examinations, one covering Calculus AB topics and the other covering Calculus BC topics, were administered to 7800 Calculus AB students and 2900 Calculus BC students, respectively. Each of the thirty-item tests contained two sections. Section I consisted of twenty items that did not require the use of calculators. Section II consisted of ten items that required calculators for their solutions. Section I items were designed to be equivalent in content and complexity to items in the current AB and BC Advanced Placement examinations. As part of the study, students also completed questionnaires about their experience with calculators, and classroom teachers described their calculator-use policy for schoolwork and homework.

Three questions framed the design of the study:

1. How does the opportunity to use calculators affect the statistical properties of individual items and the examination as a whole?
2. How does the opportunity to use calculators affect student performance on the examination as a whole?
3. Is it possible to prepare sound calculator-active items, that is, items requiring students to use calculators in order to solve the problems posed? (Jones 1991, pp. 24–26)

Findings of the study, related to the statistical properties of individual items and to the examination as a whole, showed that students using graphing calculators outperformed students using scientific calculators on several items. Although the reliability of both examinations (AB and BC) was lower when students were using calculators, using the calculator appears to produce changes in the ability of certain items to discriminate more able from less able students.

With regard to the effect of calculator use on student performance, the study indicated that using calculators resulted in better performance on Section I, noncalculator, items. Further, the type of calculator used substantially affected performance. Students using more sophisticated calculators performed better on both sections of the test. Although males performed significantly better than females in this study, it should be noted that males also reported using advanced calculators for classwork and homework more often than females did.

Finally, on the basis of the results of the study, the College Board concluded that designing calculator-active items is very difficult, particularly those that effectively discriminate more able from less able students. In general, calculator-active items are more difficult for females than for males, and most of the calculator-active items are easier for students using graphing rather than scientific calculators.

In the fall of 1990, the AP Calculus Development Committee, members of the Mathematical Sciences Advisory Committee, and other consultants reviewed the AP Calculus Calculator Impact Study. With the endorsement of the College Board's Council on College-Level Services, these groups recommended that—

> beginning in May 1993, both Advanced Placement Calculus Examinations (AB and BC) will contain questions for which a scientific calculator is necessary or advantageous. At that time, students will be expected to bring scientific calculators (from an approved list) to the examination. (Jones 1991, p. 28)

The use of scientific calculators on AP examinations mirrors the introduction of new topics in the Advanced Placement Calculus AB and BC courses. Among the topics that allow for greater use of technology that are being introduced in the AB syllabus are Newton's method and approximations to the definite integral using rectangles or trapezoids; in the Calculus BC syllabus, Simpson's rule and applications of work are being introduced.

The College Board has also identified, as a major goal, the use of graphing calculators on the Calculus AB and BC Advanced Placement examinations by May 1995. The particular content and presentation of the items on the AP examinations will reflect the availability of technology at that time. By the mid-to-late 1990s, graphing calculators most probably will have more capabilities (e.g., solve key, root finder, numerical differentiation, and numerical integration) than the models currently in common school use.

Scholastic Aptitude Test (SAT-I)

In the fall of 1990, the College Board trustees announced their approval of significant changes to the SAT (College Entrance Examination Board 1990). These changes resulted from a three-year research effort to identify, study, and evaluate modifications that would make the test more useful to students, to high schools, and to colleges. Major goals of this project were to update the test to take into account recent changes in education, advances in learning and curriculum theory, and new approaches to test design and administration.

Early in the investigation of possible changes to the SAT, advisory groups recommended the development of an SAT calculator policy that would be in accord with calculator-use practices described in the NCTM *Curriculum and Evaluation Standards for School Mathematics* (NCTM 1989). Also explored was the inclusion of questions for which students would be required to produce their own answers (and grid those answers on a special answer sheet) rather than choosing among multiple-choice alternatives. The inclusion of some questions in this format will be implemented in spring 1994.

In 1989, the College Board conducted a small-scale feasibility study of calculator use on prototype Scholastic Aptitude Test—Mathematics questions (Braswell 1990). Some questions were taken from previously administered SAT multiple-choice sections, whereas others required students to produce and record their own answers. None of the items was calculator-active (i.e., required a calculator for the solution).

Overall, there was no significant difference in performance for students who were randomly assigned calculators compared with students who did not have calculators. On the multiple-choice items specifically, there was no significant difference in performance between the two groups. However, students who had calculators did slightly less well on quantitative comparison items and slightly better on items that required student-produced (grid-in) responses.

Since the emphasis in the mathematics portion of the SAT is on problem solving and reasoning rather than on computation and advanced high school mathematics, the SAT will not include calculator-active items. However, beginning in 1994, students will be permitted (but not required) to use calculators on the examination.

Among the other recommendations that resulted from the review of the SAT was that of more closely linking the current SAT (verbal and mathematical) with the achievement tests. As a result, the Admissions Testing Program, composed of the SAT and the achievement tests, is being restructured as SAT-I: Reasoning Tests and SAT-II: Subject Tests. The SAT-II: Subject Tests will be based on existing achievement tests, with a number of revisions and new subjects. The first new SAT-II: Subject Test was the Mathematics Level IIC, which requires calculators and was administered for the first time in June 1991.

Mathematics Level IIC Achievement Test

This test assesses the same content as the Mathematics Level II Achievement Test, including trigonometry and elementary functions. However, the required use of calculators permits numerical data in test items to be more realistic. For example, angles in trigonometry items no longer have to be limited to 0°, 30°, 45°, 60°, and 90°. Students choosing to take this alternative version of the Level II test need to have had extensive experience with calculators.

The development of the Level IIC test began in 1986. Calculator-inactive, calculator-neutral, and calculator-active test items were written (fig. 20.1).

Among the challenges faced by the test development committee was the need to arrive at the correct balance of calculator activity so that the Level IIC test would not be drastically different from the noncalculator Level II test. Maintaining equivalent forms of the two tests was necessary for several

Calculator Inactive problems are those for which there is no advantage (perhaps even a disadvantage) to using a calculator:

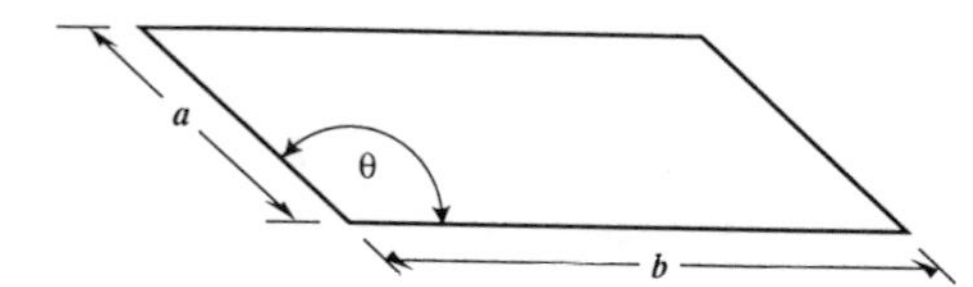

The area of the parallelogram in the figure above is:

(A) ab (B) $ab \cos \theta$ (C) $ab \sin \theta$ (D) $ab \tan \theta$ (E) $a^2 + b^2 - 2ab \cos \theta$

Calculator Neutral problems can be solved without a calculator, but a calculator *may* be useful:

$$\sin\left(\text{Arcsin } \frac{1}{10}\right) =$$

(A) 0 (B) $\frac{1}{10}$ (C) $\frac{1}{9}$ (D) $\frac{9}{10}$ (E) 1

Calculator Active problems require the use of calculators for their solutions:

The diameter and height of a right circular cylinder are equal.
If the volume of the cylinder is 2, what is the height of the cylinder?

(A) 1.37 (B) 1.08 (C) 0.86 (D) 0.80 (E) 0.68

Fig. 20.1

reasons. First, although calculator use is being widely advocated, the mathematics curriculum has not yet changed to take advantage of this use. The College Board and Educational Testing Service (ETS) develop tests that reflect existing curricula, without leading them or lagging behind. In order to remain updated on curriculum innovations and instructional issues, the College Board and ETS, through their various committees and surveys, maintain close contact with college and high school faculty. New topics are pretested and added to the examinations when their instruction is widespread.

Second, high school and college faculty are familiar with the current Level II Achievement Test and have used it with confidence. Therefore, changes in the test must be made gradually over time, so that this confidence is not eroded.

Third, the Level II Achievement Test has always been largely concept-based; the computational load is minimal. Increasing computation solely because of the availability of calculators was not deemed to be in the best interests of the students. Thus, only calculator items that measure concepts that cannot readily be taught or tested without the use of calculators are included. In the Level IIC test first administered in 1991, 60 percent of the questions were calculator-neutral or calculator-active, and 40 percent were calculator-inactive.

A major difficulty in the development of calculator-active problems was the preparation of distractors for multiple-choice responses. Pilot tests of calculator items showed that students often pressed wrong buttons or had their calculators in the wrong mode (i.e., degree versus radian). Test developers had to decide whether or not to include answer options that resulted from these types of errors. For example, when assessing the concept of the inverse of a trigonometric function, should reciprocals be included among the answer choices, since knowing they are not correct is central to an understanding of the problem? The developers recommended that test directions caution students to set their calculators in the correct mode.

Test questions were finalized and pretested with high school students in 1988. A full prototype test was administered in the fall of 1989 to approximately 2000 entering AP Calculus students. This population was selected because it closely represented the degree of academic preparedness of Mathematics Achievement Test Level II students.

The prototype test questions elicited calculator use very accurately. That is, calculators were most often used on calculator-active questions. The field test demonstrated that the content validity was not eroded by the use of calculators.

As a part of the field test, students completed questionnaires about their classroom experience, the types of calculators they used on the test, and, for each question, whether or not they used their calculators. From responses to the questionnaires as well as item analyses and other statistical data collected from the administration of the prototype, test developers determined that students who used graphing calculators had no achievement advantage over students who used nongraphing calculators. This situation may change as graphing calculators become more readily available to students.

The pilot study also showed that students who used calculators regularly and were familiar with the particular make and model calculator they used on the test performed significantly better than students who did not regularly use calculators in class or who used different calculators in class than on the test. Therefore, the College Board decided that during the actual nationwide administration of the Mathematics Level IIC Achievement Test, students would bring their own calculators. In 1991, the calculators that were allowed were the eight most commonly used calculators, as reported by students during the field test.

CONCLUSION

As new technologies become available, the study and investigation of mathematical concepts and topics previously not accessible to students will become possible. Curriculum will change. Pedagogy will change. New curricula and pedagogy will necessitate changes not only in the content of standardized tests but also in the types of technology permitted for use by test takers. The College Board and other test developers will need to establish policies on the use of calculators that are fair to all students, that are in concert with classroom practice, and that are consistent with the recommendations of the professional organizations that are setting curricular, teaching, and testing standards for our nation's youth.

REFERENCES

American Association for the Advancement of Science. *Science for All Americans: A Project 2061 Report on Literacy Goals in Science, Mathematics and Technology.* Washington, D.C.: The Association, 1989.

Braswell, James. *Feasibility Study of Calculator Use on the Scholastic Aptitude Test.* Draft report. Princeton, N.J.: Educational Testing Service, 1990.

College Entrance Examination Board. *Academic Preparation for College: What Students Need to Know and Be Able to Do.* New York: The Board, 1983.

———. *Academic Preparation in Mathematics: Teaching for Transition from High School to College.* New York: The Board, 1985.

———. *Background on the New SAT-I and SAT-II.* New York: The Board, 1990.

Jones, Chancey. *The Use of Calculators in College Board Mathematics Examinations in the 1990's.* Draft report. Princeton, N.J.: Educational Testing Service, 1991.

Kenelly, John W., ed. *The Use of Calculators in the Standardized Testing of Mathematics.* New York: College Entrance Examination Board, 1989.

Mathematical Sciences Education Board, National Research Council. *Reshaping School Mathematics—a Philosophy and Framework for Curriculum.* Washington, D.C.: National Academy Press, 1990.

Morgan, Rick, and Joe Stevens. *Experimental Study of the Effects of Calculator Use on the Advanced Placement Calculus Examinations.* Research report no. 91-5. Princeton, N.J.: Educational Testing Service, 1991.

National Council of Teachers of Mathematics. "Calculators in the Mathematics Classroom." Position statement. Reston, Va.: The Council, 1986.

———. *Curriculum and Evaluation Standards for School Mathematics.* Reston, Va.: The Council, 1989.

Part 5

21

Calculators Add Up to Math Magic in the Classroom

Karen W. Finley

FIVE years ago I started teaching mathematics at the fourth- and fifth-grade levels. My first year was spent teaching in a very traditional way—follow the textbook, drill the number facts, try to make place value make sense, suffer through fractions (the students *and* I), and if I ever got anywhere near the introduction of decimals, it was definitely time to celebrate! The word *manipulative* was not even part of my vocabulary, much less my classroom practice. Fortunately, I have always had a love of mathematics. I say *fortunately* because I wanted so much to pass this love on to my students that I did try to be somewhat innovative in my approach to the teaching of what most students consider a painful subject. I knew that my students didn't really hate *math*. They hated the frustration, confusion, lack of purpose, and general sense of helplessness that they had come to associate with that four-letter word (unless, of course, they were among the chosen few who happened to be born with that special talent that made them "good at math"). My goal was to see my students' faces light up with excitement and anticipation whenever I mentioned that four-letter word.

GETTING STARTED

During my second year I was introduced to manipulatives through a program initiated by the school administration to help teachers approach the teaching of mathematics in a more meaningful and effective way. My classroom began to bustle with activity during math as more and more tools for learning were introduced to the children. One of these many tools was the calculator.

During one of our faculty seminars, the question had been posed, "Have you ever considered allowing students to use calculators in class?" Of course none of us ever had. Children sometimes brought them to school to play around with just as they would any other "toy" and would sometimes joke about using them during class, but that was only a joke, wasn't it? I decided to try using calculators in my class. I felt that I wanted to explore the use of calculators as a much-needed tool for problem solving and critical thinking.

The next day, I composed a letter to be sent home to the parents of all my students explaining that I was going to allow the children to use calculators in class and with their homework under very specific circumstances. They were only to use the calculators if they were doing word problems. I explained that calculators are tools that help us with the computational part of problem solving and that they cannot solve the problem for us. The child must know first what he or she needs to do to solve the problem before entering it into the calculator for computation; otherwise, the calculator is useless. It is much more important that they know that they must multiply 4×7 in order to solve the problem than that they know that $4 \times 7 = 28$. I assured the parents that they should not worry about their children's computational proficiency suffering in any way, because I would make sure that the basic number facts would still be drilled and tested on a regular basis.

Needless to say, I expected my presentation to the children on the use of calculators and the rules they would have to abide by would be met with no less than three rounds of "For she's a jolly good fellow. . . ." Was I wrong! The faces that stared at me as I passed out the letters to be taken home were full of apprehension, amazement, and mistrust. They weren't quite sure if they should be happy or not. I could see in their faces that they weren't sure if they could trust me as a teacher if I was actually going to allow them to "cheat." After all, it was my *job* to *teach* them and to make sure they didn't cheat. So, the greatest resistance came from the one group I least expected it from—the students.

I solved that problem by saying that we would use the calculators on a trial basis. Each student would have the option of using the calculator or not, and if things didn't work out well after a month or so, then we would reevaluate the situation. This made the students feel a lot better. I think they felt a little less threatened and out of control in the face of this radical proposal.

THE TRANSITION

As the days passed, calculators began to appear in our classroom sporadically. The students began to use them, at first somewhat reluctantly, and the easy transition period I had expected was very slow in coming. After a couple of weeks, I asked the children to write a short paragraph expressing their feelings about the use of the calculators. The response was almost unanimous. They felt that they were somehow cheating and that this was not the proper way to do math. The calculator was doing the work for them, which wasn't fair to them.

We discussed this issue in class, and I ended the discussion by proposing a contest between them and the calculator. I took out a calculator, placed it on the desk, and asked the students to take out paper and pencil.

"I will write a multiplication problem on the board and we'll see who can solve it faster, you or the calculator. Who do think will win?"

They all laughed and said that of course the calculator would win. I asked them to give it a try anyway, wrote the problem on the board, and gave them the "ready, set, go" signal. The students were busy calculating while the calculator just sat on the desk. The first student to get the answer jumped up and called it out, at which time she was pronounced the winner!

"You beat the calculator!" I said. "Look! The calculator hasn't even started yet!"

They looked at the calculator and laughed. Finally! Faces of enlightenment!

"We beat the calculator because the calculator can't work by itself," they said. "The calculator needs us to tell it what to do. The calculator does only what we tell it to do."

"That's right," I said. "And suppose you enter the wrong numbers by mistake. Will the calculator correct it for you?"

"No!"

"So that means that you have to pay close attention to what you ask the calculator to do as well as to the answer the calculator gives you, because it is possible for the calculator to be wrong."

That lesson took place four years ago. The repercussions have been numerous and unending. Today my classroom is equipped with calculators for each student, and the calculators are in constant use. The children are very comfortable with them and are discriminating users of the calculators. They realize that the calculator has limitations and is therefore not useful in all situations. However, they have not yet realized the calculator's full potential. Nor have I—I am still learning with and through my students.

POSITIVE RESULTS

I have seen many changes take place in my mathematics class and in the school—changes that I attribute directly to the use of calculators. When the students became aware of the fact that they did indeed have to check the calculator's response for accuracy, it forced them to check the logic of their answer. All those years of telling them to ask themselves if their answer made sense went in one ear and out the other until they started using the calculator.

The students have also become much more proficient at estimating and predicting outcomes. This is because they are no longer bogged down by tedious computations and are free to spend more time thinking about the process they want to use and what they expect to happen as a result. When the calculator does not give them the result they expect, the students want to know why. They are much more willing to reevaluate their own thinking, retrace their steps, and try different manipulations until they can make the process and results reasonable. The calculator has allowed students to follow their thought processes through to completion without unnecessary interruptions.

One especially gratifying incident occurred in that first year of calculator use. I found one of my students, who had a particularly hard time with mathematics, working very intently on a problem and using the calculator to try out his different theories. It was rewarding enough that he had progressed to the point of even having a theory and that he was willing to try it out on his own without asking for help first. But my true reward came when he finally approached me, with calculator in hand, and said, "Is this right?" Before I could say a word, he looked at the calculator again and said, "No, that's not right. Something is wrong. Let me try again. I think I know what I did wrong." He didn't even wait to hear my response or look up to see the big smile on my face. He was too intent on solving his problem. At the beginning of the year he would never have got past the first simple computation using paper and pencil before he would have been totally frustrated and unable or unwilling to pursue any kind of logic.

Calculators have been wonderful as a tool for building confidence and self-esteem in students. Students love to be right and to feel that they have the ability to reason something through. The calculator affords them many opportunities to be correct and to receive instant gratification. When they are using paper and pencil and make a computational error, they are more often penalized for the error in computation than praised for correct reasoning. Even if they are praised for their reasoning, the praise is somewhat tainted by the fact that the answer was not exactly right. The calculator helps the teacher in getting the point across that reasoning is the most important step, and it allows the student to feel the great satisfaction that accompanies success.

EMPHASIS ON NEW SKILLS

The students in my class now do a lot of mental arithmetic and do it faster because they have become accustomed to fast answers from the calculator. This has an added bonus of improving their number-fact skills. They tend to remember the multiplication tables better because they need them more. They also *see* them more because of the calculator. They might enter 9×7 into their calculator many times in forty minutes and see the answer 63 over and over again. Mentally or verbally, the students are usually saying each computation to themselves as they enter it—a definite method of drilling. Consequently, I spend less class time drilling number facts and more time looking for patterns and the relationships of numbers.

Many of my lessons are now built around the calculator simply because a student was faced with the dilemma of how to enter a problem into the calculator. For example, one student went through the following process:

Question: 20% of 60 = _______

Student's work: 1/5 of 60 = _______

5 × _______ = 60

At this point he picked up the calculator and started to enter 5 × . . ., but then he realized he couldn't ask the calculator "5 times what equals 60." He had to figure out another way to ask the same question. He thought it through and came up with 60 ÷ 5 = 12. The calculator helped him understand the relationship of products and factors in a way that was very meaningful to him because it came from a need to know.

Another student wanted to use the calculator with fractions. His question was, "Is there a way to enter fractions into the calculator?" He felt certain that there must be because fractions and decimals are related. If we can enter decimals, we must be able to enter fractions. I told him that there was a way and to see if he could find it. I let him work on it for a while before I gave him the big hint that he should remember what the line between the numerator and the denominator of a fraction really means. Bingo! 3/4 means 3 divided by 4. The result on the calculator is a decimal. He was quite impressed and spent some time changing fractions to decimals.

OBSERVATIONS

I have found that children who are exposed to these calculator experiences find going from the concrete to the abstract a much easier transition. The curriculum must keep pace with this acceleration. What we once thought was beyond the capabilities of a certain age group is no longer out of reach. The more children are allowed to explore and question, the more prepared we must be to guide them to the next step. I was once asked by a student what happens when we have a number that has too many digits for the calculator to display. Another student immediately offered the possible solution of using a computer instead of a calculator. This was the perfect opportunity to explore the area of scientific notation.

The calculator is just one small tool, but I feel that it has had a dramatic effect on how, when, and what children learn. Calculators are used throughout our school now, along with many other materials, and each year I see students entering class a step (or two or three) ahead of the year before. I find it very challenging, exciting, and rewarding!

22

Implementing Calculators in Middle School Mathematics: Impact on Teaching and Learning

Gary G. Bitter
Mary M. Hatfield

IN VIEW of the evidence indicating the effectiveness of calculators in mathematics instruction and given the endorsements of calculator use by virtually every mathematics education organization, the failure of schools to exploit calculators in mathematics instruction is puzzling. Calculators have become ubiquitous, but regular classroom use is sporadic.

DESCRIPTION OF THE PROJECT

This article describes an Arizona project that combined the resources of school district personnel, university mathematics educators, and private industry to improve the use of calculators in mathematics instruction. The project resulted from the efforts of an individual teacher assuming a leadership role as a change agent. The teacher submitted a request to the school principal that calculators be provided on a daily basis for all students. The district's school board approved the implementation project and purchased Texas Instruments Math Explorer calculators for all students to use beginning in the fall of 1988. This calculator was chosen because of its fraction/decimal function keys—functions that address a large component of mathematics instruction in the middle grades. The principal sought educational and research support for the project from Arizona State University mathematics educators, who subsequently requested assistance from Texas Instruments.

The setting for the project was a middle school in the metropolitan Scottsdale area attended by 580 seventh and eighth graders. Fifty-four percent of the subjects were seventh graders, and 46 percent were eighth graders. The gender distribution was 54 percent male and 46 percent female. Seventy-four percent of the subjects were white, 10 percent were Hispanic, 9 percent were Native American, 2 percent were African-American, 2 percent were Asian/Pacific Islander, and 3 percent did not report their ethnicity. The socioeconomic level of the families is middle/lower class. The student population at this school is highly transient with a turnover of approximately one-third of the student population yearly. This school consistently has scores on the Iowa Test of Basic Skills (ITBS) mathematics subtests and the district's Mathematics Inventory Test that are below the rest of the district. Furthermore, one-third of the incoming seventh graders qualify for Chapter 1 services in reading and mathematics.

The main goals of the study were to determine (1) the effects of a calculator-enhanced mathematics curriculum on seventh- and eighth-grade students' mathematics achievement, (2) the effects of a calculator-enhanced mathematics curriculum on students', teachers', and parents' attitudes about using the calculator for mathematics, and (3) the implementation effects of in-service training and support on teachers' instructional uses of the calculator. The pilot year of the project was the 1988–89 school year. During the 1989–90 school year, the second year of the project, revised assessment instruments were used and implementation procedures were refined.

The project staff consisted of members of the mathematics education faculty and a graduate assistant at Arizona State University. The school principal and the lead teacher (who initiated the project) played a pivotal role in the success of the project.

The framework of the implementation was to furnish each student with a calculator for use throughout the year in all classes and to provide extensive, continuous in-service sessions for participating mathematics teachers. Calculators were allowed to remain with the students at all times. They could be taken home, used in any class, and used on tests.

PROJECT RESULTS

Results from the analysis of quantitative and qualitative data collected during the project tended to reaffirm the benefits of calculators for student performance and attitudes. Students performed significantly better on three mathematics subtests of the Iowa Tests of Basic Skills (ITBS) when they were allowed to use a calculator. The attitudes of students, teachers, and parents as measured by the revised Pocket Calculator Attitude Scale (Bitter 1980) were almost universally favorable, yet responses from each of these groups revealed an awareness of the limitations and potential dangers of unrestricted or poorly planned calculator use. Other findings showed that girls' performance improved more than boys' with a calculator, that girls experienced a greater feeling of empowerment, and that how a teacher integrates calculator use in his or her classroom curriculum is a critical factor in student achievement and attitude.

IMPLICATIONS FOR IMPLEMENTATION

The project revealed a number of strategies likely to lead to successful incorporation of calculators in mathematics programs.

Setting the Stage

When curricular change and reform efforts, such as the implementation of calculators, are required, our experience suggests that it is necessary to begin with careful, well-coordinated steps to ensure broad-based support from participating school personnel. Our experience also indicates that an implementation plan aimed at making widespread change in the teaching and learning of mathematics receives better overall acceptance and cooperation when it is a school-based decision. Including teachers at the school level in the decision making increases the likelihood of the long-term success of the program. The locus of change is the individual school.

Our experience suggests that it is beneficial for teachers to become familiar with reports and documents such as the *Curriculum and Evaluation Standards for School Mathematics* (National Council of Teachers of Mathematics 1989) that provide a framework for using calculators in mathematics instruction. Additionally, from our experience, the investigation and discussion of calculator research results is a useful method of improving and refining calculator integration methods in the classroom. The encouragement of free, open discussion of teachers' concerns or lack of commitment, as well as conditions for success, possible obstacles, and other issues that may be encountered, can increase the understanding and acceptance of the program. For the same reasons, our findings indicate that administrators, too, should be encouraged to participate and demonstrate an active, interested role in the project and that communication with parents, in advance, about the educational effectiveness of calculators and details of the proposed implementation project can build understanding and acceptance. We provided parents with frequent information about how calculators are being used in school and with calculator activities that could be completed at home.

Professional Development

Many educational innovations are only marginally effective in promoting lasting changes in teachers' behavior and instructional practices. Uncertainty about expectations, resources, pedagogy, and content knowledge are some of the factors that influence the implementation of any new program. The way in which teachers themselves have been taught has a profound effect on how they view what constitutes appropriate teaching and learning. In the NCTM *Professional Standards for Teaching Mathematics* (National Council of Teachers of Mathematics 1991), the education of teachers of mathematics is called an on-going process.

> [Teachers'] growth is deeply embedded in their philosophies of learning, their attitudes and beliefs about learners and mathematics, and their willingness to make changes in how and what they teach. Their growth is also affected by numerous external agents including school administrators, educational policy-makers, college and university faculty, parents, and the students themselves. (p. 125)

To address some of the needs of the continuing education of teachers, the implementation of the project was designed to give multifaceted support. Throughout the pilot year, participating teachers and their principal attended weekly hour-long in-service sessions before school. In the project's second year, in addition to individual planning times, the teachers were assigned to a common planning time with other teachers of the same subject area. The purpose was to provide time to collaborate with peers, to encourage interaction among teachers, and to allow the professional development of teachers to be conducted during the duty day. The project's graduate assistant was on site with participating teachers providing training, assisting in teaching lessons, and observing project teachers regularly as part of the evaluation of teaching.

We found that it is important to include in-service experiences that serve as models of what it means to teach and learn mathematics. In the Scottsdale calculator project, the implementation design focused on a comprehensive, ongoing in-service support system (fig. 22.1).

The implementation of a successful staff development program requires long-term support, appropriate activities that engage teachers, planning and needs identification by participating teachers, and many forms of administrative support.

Administrative Leadership

We found administrative leadership to be a key component in implementing an effective calculator program and in convincing others of the educational value of calculators. Our experience suggests that administrators should be instrumental in setting up a calculator policy that is understood by teachers, students, and parents. We found that consistent promulgation and application of the policy, which may differ at different school levels and which may be based on any of several existing models, such as the calculator policy advocated by the NCTM, is an important contributing factor to the overall effectiveness of the project. The project was grounded

Calculator Implementation Support System

- Mathematics learning experiences were constructed to engage teachers in a direct, dynamic way as they participated as learners in a learning environment.
- Teachers worked through sample calculator lessons in a problem-solving environment using a collaborative learning model.
- Sample lessons integrating the use of the calculator into the existing mathematics curriculum were developed and modeled by the project's graduate research assistant.
- Projects and activities were suggested and experienced.
- Guidelines for teaching with a calculator were generated and adopted.
- Activities were designed to encourage teachers to help students learn to select the most useful means for working problems: calculators, conventional computation, or mental computation.
- Teachers exchanged ideas, shared teaching strategies, and discussed problems and solutions related to calculator use.
- Participating teachers were encouraged to request the content-specific training they wished the project staff to provide.
- Project staff observed the teachers regularly and provided counsel in improving instruction.
- Project staff were available to teach a class so teachers could visit one another's classes to benefit from another teacher's approach.
- Frequent discussions among the project staff and teachers aimed at encouraging teachers to engage in the process of self-reflection and self-analysis as part of an effort to improve instruction.
- Teachers were encouraged to attend state mathematics meetings, and monies were provided to attend the annual NCTM convention.

Fig. 22.1

in the early, continual, proactive support of the school principal, who encouraged teacher involvement and promoted the project's ideals to the district school board.

Administrators should recognize that calculator implementation involves a long-range design of planning, implementing, and evaluating the program. To have an impact on instructional change, calculator implementation should, according to our findings over a two-year period, be part of an ongoing program of equipment acquisition and curriculum development. Records maintained during the term of the project on the wear, loss, and breakage of calculators issued to each student suggest that administrators should budget for repair or replacement on a three-to-five-year cycle; concomitantly, our experience suggests that it would be useful and appropriate for administrators to remain informed of advances in educational calculator technology and other curriculum resources and to plan for updating where appropriate. Technology companies should collaborate with administrators in planning implementation strategies that are economically feasible and educationally sound. To be successful, implementation requires a solid commitment of resources, funding, and time.

Involving Parents

From our experiences, we believe that parents should be informed and involved when curriculum restructuring is planned, especially when the change may involve new technologies or parental expenses. Most of today's parents did not use calculators when they attended school, and they may have beliefs and perceptions that ignore the potential of the calculator as a tool to enhance mathematics instruction. Educating parents that no evidence exists to suggest that the availability of calculators makes students dependent on them constitutes an important component of the implementation plan.

Schools should prepare for, and may face, some resistance when parents are asked to replace a calculator if it is lost or broken. Data on parental attitude show that about 88 percent of the project's parents were in favor of making the calculator a regular part of the school curriculum. The success of the project was enhanced by involving parents in the project by several forms of communication, including informational letters, questionnaires, and meetings at school. Our experience shows that involved parents will be supportive of a well-prepared, systematic approach to integrating calculators into their child's mathematics curriculum.

Ensuring Equity and Access

Appropriate calculators should be available to all students across all curriculum areas. Guaranteeing such equitable access is becoming more realistic as calculators become less expensive, use solar power, and operate more uniformly. Questions still arise today, however, on the logistics of ensuring that students have access to an appropriate calculator. The project results indicate that access is not an issue, since parents and students report that calculators are available in 97 percent of the students' homes. The data suggest that four-function calculators are ubiquitous; therefore, it is reasonable to expect students to bring calculators to school in the same manner as they bring pencils and paper. Since this project involved a specialized calculator, one calculator was assigned to each student like textbooks and other school property, and parents signed an agreement to replace it if lost or stolen.

Using Calculators in the Classroom

Many factors can facilitate calculator implementation in the classroom. A lead teacher can be a strong influence in calculator adoption; in addition to initiating the project, our project's lead teacher supported other teachers in their ongoing efforts at curriculum integration. The concept of a lead teacher can be extended to a "lead school" or even a "lead district," where ideas can be tried, evaluated, demonstrated, shared, promulgated, and supported.

Calculators by themselves do not make a lesson or a curriculum. Before changing classroom activities, most teachers want effective resource materials that can augment their lessons. Indeed, in the absence of a calculator-based mathematics curriculum, the quality of available resource materials is a primary factor in teachers' use of calculator technology. From the standpoint of instructional design, technology is not as important as the body of instruction in which it is embedded. Even though textbooks have not yet fully integrated calculator pedagogy, many resources do exist and are available from publishers and calculator manufacturers. Our project used a resource book (Bitter and Mikesell 1990) for curriculum-based ideas and activities. Teachers and administrators can check with journals, ancillary materials, libraries, other schools and districts, university educators, and technology companies in obtaining similar resources. Our calculator study suggested a number of practical considerations when using calculators in the classroom (fig. 22.2).

General Tips for Classroom Calculator Implementation

- Provide students with information on the care and handling of calculators.
- Introduce calculators after students have been involved with hands-on manipulative activities.
- Allow free time with the calculators so the novelty wears off.
- Do not "teach" the calculator; rather, have students discover the calculator's features.
- Have students explore and discover their own algorithms, using the calculator to verify results.
- Encourage the estimation of answers before using the calculator.
- Use real-world data and real-world problems; modify problems to involve students and to ensure curriculum relevance.
- Emphasize activities and projects that involve higher-order thinking.
- Use the calculator as a tool to teach and learn mathematics rather than as the object of instruction.
- Encourage students to evaluate the appropriateness of using the calculator instead of estimation or mental arithmetic.
- Allow students to work in pairs or groups when appropriate.
- Encourage unconstrained and creative approaches to mathematics problems.

Fig. 22.2

Testing

Perhaps the major factor that discourages calculator implementation is that most standardized test do not currently allow calculators. Understandably, most teachers and school districts are concerned with standardized test results. From an instructional design point of view, calculators should not interfere with assessment. For instance, calculators would be inappropriate for items designed to measure a specific paper-and-pencil skill. However, using calculators on noncomputational test items helps ensure that students are not being penalized for weak computational skills (Reys and Reys 1987). The calculator helps isolate particular higher-order outcomes addressed by conceptual or problem-solving test items.

CONCLUSION

Recent calls for educational reform consistently advocate an inquiry-oriented learning environment that promotes the development of students' mathematical power. The calculator can be used effectively in establishing such an environment if it is used as a tool for mathematical explorations and investigations. As happens with almost any technological or curricular change, the use of calculators or calculator-enhanced mathematics instruction generates important implementation issues. The Scottsdale calculator project suggests some strategies to help teachers, administrators, and school districts integrate calculators into mathematics instruction. Implementation, to be successful, requires multifaceted support, a solid commitment of resources, an appropriate and comprehensive professional development program, and a systematic long-term commitment to appropriate change. We believe that when these features are part of a calculator-based innovation, chances for success are very high. In our project, we found an improvement in student performance and attitude, and we found that parents and teachers will support a well-planned calculator-integrated curriculum.

REFERENCES

Bitter, Gary G. "Calculator Teacher Attitudes Improved through Inservice Education." *School Science and Mathematics* 80 (April 1980): 323–26.

Bitter, Gary G., Mary M. Hatfield, and Nancy T. Edwards. *Mathematics Methods for the Elementary and Middle School: A Comprehensive Approach*. Needham Heights, Mass.: Allyn & Bacon, 1989.

Bitter, Gary G., and Jerald L. Mikesell. *Using the Math Explorer™ Calculator: A Sourcebook for Teachers*. New York: Addison-Wesley, 1990.

National Council of Teachers of Mathematics. *Curriculum and Evaluation Standards for School Mathematics*. Reston, Va.: The Council, 1989.

———. *Professional Standards for Teaching Mathematics*. Reston, Va.: The Council, 1991.

Reys, Barbara J., and Robert E. Reys. "Calculators in the Classroom: How Can We Make It Happen?" *Arithmetic Teacher* 34 (February 1987): 12–14.

23

Implementing Calculators in a District Mathematics Program: Three Vignettes

Douglas B. Super

Educational change is a complex, complicated process which involves wanting and learning to do something new. (Fullan 1982)

SUCCESSFULLY introducing the use of calculators into a school district is sometimes more difficult than it would seem. Despite strong arguments from mathematics educators and researchers favoring calculator use, resistance can be encountered from the community, parents, school board members, district and school administrators, and teachers. Such resistance can be subtle or insistent, and it is certainly damaging unless thoughtful contingency planning is undertaken by the coordinators of an implementation. It is vital to analyze goals and objectives and to consider possible obstacles and opportunities before melding suitable strategies into a coherent implementation plan (Price and Gawronski 1981). This article reviews three very different large-scale implementations of calculators in two different school districts.

SUBURBAN DISTRICT: ELEMENTARY SCHOOLS

In the 1980s, prompted by the *Agenda for Action* (National Council of Teachers of Mathematics 1980) statement on technology, a K–12 consultant for mathematics and computer education in a suburban school district with an enrollment of 18 000 students began informally analyzing the impediments to technology-related change in school mathematics. Obstacles to wide-scale implementation of calculator use seemed to abound: misunderstandings by parents and the general public were evident from casual conversation, and school board members were unlikely to approve a sizeable expenditure on such a controversial item. There was even a mixed reaction from teachers. What seemed to be needed was some kind of proof—not proof from an erudite monograph or inconclusive experiences from a distant district but home-grown proof in the local schools with data that were substantial and generalizable—that calculators can have positive impact on student performance and attitude.

> *Assumptions about introducing a change often do not adequately consider the concerns of teachers.* (Hall and Loucks 1978; Austrom et al. 1989)

The possibility of a large-scale research project involving many seventh-grade teachers in the district was discussed by the consultant and his supervisor. The short-term plan was to conduct a pilot project and then push for implementation across the seventh grade if the pilot was successful. The long-range plan was to move down the grade levels one year at a time as teachers at adjacent grade levels became convinced of the benefits of calculator use. Because funds to equip each seventh-grade classroom in the district were lacking, a decision was made to involve a sizeable control group, so that most teachers at the grade level felt part of the research. Fortunately, a respected faculty member from a local university became a coinvestigator, bringing with him the prestige of university research as well as funding for the development of materials, classroom visitation, and assistance with testing and evaluation.

The study lasted one entire school year. Three groups were established: a control group, a group provided with problem-solving in-service programs and materials, and a group provided with both calculator and problem-solving in-service programs and materials. The research was successful, adding to the knowledge base on the effects of calculators on student performance. For example, students in the calculator group performed as well or better on standardized, criterion-referenced, paper-and-pencil tests and better on problem-solving measures that permitted the use of calculators (Szetela and Super 1987). More important for implementation than the statistical results were the lessons learned about the power of commitment by teachers and about the effectiveness of classroom-based research in building acceptance and knowledge of an innovation. Similarly, interest was raised for teachers in the control groups and for other grade-level teachers in the schools with calculator pilots.

> *Teacher characteristics such as attitudes, beliefs, experiences, and abilities have a major impact on the outcome of planned change.* (Fullan 1982; Berman and McLaughlin 1978; Austrom et al. 1989)

Several practical lessons were learned from the implementation:

1. If possible, each classroom should have its own set of calculators or-

ganized for easy access by students. It is just too likely that teachers will minimize the use of calculators when access is limited.

2. It is unreasonable to expect teachers to adapt *on the fly* curriculum materials or textbooks to incorporate calculator use. Resources need to be gathered and correlated to texts. Even though calculator use encourages situational investigations that arise naturally in the classroom, these investigations are not enough.

3. One-shot workshops do not change teacher beliefs; isolated calculator units do not lead to a rethinking of the role of skill development. From extended use by students with a variety of topics, more and more teachers become convinced that calculators are a valuable tool and become skilled at making appropriate instructional decisions regarding calculator use in a mathematics program.

> *Educational changes benefit enormously from parents' understanding and participation, and indeed will probably fail to be implemented if parents are ignored or bypassed.* (Fullan 1982; Austrom et al. 1989)

During the year of research with seventh-grade teachers, numerous school-based calculator workshops were held for parents. By focusing on challenging exploratory topics and stressing the need for continued basic-skill development as well as increased problem solving, the workshops satisfied all but the most adamant parent skeptics. However, it was noted that the concerns raised during the question period were frequently the same, despite the fact that most were addressed during the presentation itself. As a result, when the provincial mathematics association commissioned a pamphlet for teachers and parents on the topic of calculator use, a question-and-answer format seemed most appropriate. To everyone's surprise the pamphlet became a best-seller in the region, long after the consultant had moved to a different district (Super 1988).

URBAN DISTRICT: ELEMENTARY SCHOOLS

The second calculator implementation experience was set in a metropolitan district of more than 50 000 students with diverse ethnic and socioeconomic backgrounds. The district organization and its internal decision-making processes were far more complex than those of the suburban district discussed previously. The research approach that worked so well in the suburban district was judged to be inappropriate. Other opportunities were available, however. A newly mandated curriculum contained the reference that "all students will have access to calculators." In addition, the text resources adopted by the urban district each had extensive calculator activities in the pupils' and teachers' editions. Third, excellent supplementary resources were available, one of which was tied closely to an adopted program (Beattie and Szetela 1988). Clearly, some encouraging factors were present that had not been available in the suburban district.

Three significant obstacles, however, became quickly apparent:

1. The school-based budgeting tradition could prevent the selection of a standard calculator and the purchase of enough sets to have classroom-based calculators.
2. Negative statements by some members of the school board and by the local press caused concern and might severely hamper implementation efforts.
3. Teachers and schools were being overwhelmed by several curriculum changes and instructional improvement initiatives; introducing calculators might make only a very small ripple in a very large and turbulent pond.

With the district's director of instruction as an ally, the supervisor of academic programs began to circulate the aforementioned parent brochure to all district and school administrators. The necessary political support to forge ahead was secured at a crucial presentation to the Education Committee of the School Board, using insights gained in countless parent workshops. A personal decision was made by the supervisor to be relatively inaccessible to the press, and after several days the "issue" passed from reporters' minds.

The purchase of calculators required accessing strapped operating funds, since special budget allocations in public education often lag behind (and sometimes fail to reflect) established policy. At a second crucial meeting, this move garnered critical support from those assistant superintendents working most closely with the schools. Also, the district's seventy-five elementary principals agreed to the central coordination of their calculator purchases as part of a formal implementation plan for a revised K–7 mathematics curriculum, since the plan was developed in conjunction with a committee of their colleagues. Thus, two of the obstacles noted above were circumvented.

> *The role of the administrator in the institutionalization of an innovation is critical.* (Fullan 1982; Austrom et al. 1989)

At this point a key implementation strategy was devised to overcome the third obstacle. As a result of residual uneasiness concerning community politics and a lack of uniform acceptance or excitement by overloaded teachers, the final implementation plan allowed for the dispersion of calculator resources to schools only after the completion of a half-day, school-based training session. This was not a universally popular strategy with school administrators and the teachers' association, but it turned out to be a crucial strategy in the process of implementation. It managed to gain the attention of schools throughout a relatively unfocusable district. A staff-development team of five primary teachers and five intermediate teachers eventually presented a carefully crafted workshop in each elementary school. Staffs received not only separate training for primary and intermediate teachers but also parent brochures for every child, commercial and district-produced instructional resources, an overhead calculator, a calculator for every student in grades 5–7, and additional class sets to be shared by classrooms in grades 1–4. This allotment, which consumed the available budget, decreased access at the fourth-grade level where unclear research results were noted by some of the local critics of calculator use (Hembree and Dessart 1986).

> *Innovations that are successful combine good ideas with good implementation support.* (Fullan 1982; Austrom et al. 1989)

Despite the motivational in-service programs, misunderstandings regarding the appropriate use of calculators were noted by some members of the staff-development team in follow-up discussions. As a result, it was decided to send a one-page set of guidelines (fig. 23.1) to each elementary school teacher and administrator in the district. The guidelines were designed to meet three needs: as a reminder of appropriate instructional practices, as a reference for teachers defending the use of calculators with parents, and as reassurance to the community and school board.

After peak activity had subsided with the completion of the K–7 mathematics implementation, a questionnaire survey was conducted. The results showed—

1. widespread acceptance of the use of calculators;
2. a need for classroom-based assessment tools;
3. a strong request for additional resources by teachers in grades 1–4.

The staff-development team was reconvened to prepare suitable testing materials, and a tentative plan was devised to free up calculators for the primary level by furnishing to the intermediate grades calculators that can handle fractional operations.

URBAN DISTRICT: SECONDARY SCHOOLS

The third implementation experience was in the same urban district and involved eighteen secondary schools served by one K–12 consultant for mathematics and science. Department heads met monthly with the consul-

Guidelines for Calculator Use

INSTRUCTION

It is recommended that calculators—

1. be available for problem-solving situations;
2. be used by students to perform difficult computation related to real-life applications;
3. at times, be used to develop mathematical concepts and processes, such as place value and patterning;
4. be introduced so that special keys and functions are used appropriately;
5. not replace the need for appropriate computation skill with paper and pencil;
6. increase the need for students to have proficiency with mental math and estimation, both requiring skill with place value and basic number facts.

STUDENT EVALUATION

It is recommended—

1. that assessment of a student's computational performance take into consideration facility with mental math, estimation, and use of the calculator, as well as paper-and-pencil skills;
2. that calculators should *not* be available to students for standardized tests, provincial and district assessment tests, or tests and quizzes provided with prescribed text resources, unless stated otherwise in the test instructions.

Fig. 23.1

tant to discuss curriculum, resources, assessment results, test-item banking, and other issues. The district was in the midst of a four-year phased implementation of a revised regional curriculum and newly prescribed textbooks for the eighth through twelfth grades. The culture of the secondary schools and their approach to innovation and decision making were quite different from that of the elementary schools, and this played a major factor in formulating plans for introducing calculator use.

In the same year that the district purchased calculators for elementary schools, it was decided to purchase for eighth-grade teachers just one class set for each school (simple four-function calculators, as referred to in the adopted texts). Also, students were encouraged to purchase the same type of calculator. About two hours of in-service instruction was provided for each teacher of the eighth-grade course. Given the limited support offered, the classroom use of calculators varied widely in the junior secondary classrooms, and implementation was deemed partially successful at best. In lengthy follow-up discussions, a strategy was devised to provide more attractive resources and more comprehensive in-service programs for teachers of eleventh- and twelfth-grade courses and to use the anticipated success to revisit the eighth- through tenth-grade courses from the perspective of, and with the prestige of, advanced coursework.

A year passed with actions devoted to preimplementation at the eleventh- and twelfth-grade levels. At first, it was unclear whether scientific or graphing calculators should become the district standard. With increased support from the regional mathematics association, the promise of dropping prices, and a comparison of the calculators' educational versatility, it was decided to go with graphing calculators. At a price of twenty times a basic four-function calculator, funding became a central issue. In the spring, before the implementation of revisions in the eleventh grade, an action-research approach was adopted that incorporated several components.

Two full sets and three half-sets of graphing calculators were offered on a competitive basis to schools with teachers who were willing to participate in classroom-based action research, to assist with the production of district resources, and to provide in-service programs for peers at a district conference one year later. In addition, those teachers with the half-sets would be trained in cooperative learning techniques, and they agreed to employ this strategy in their action research. The response to this offer was restrained; exactly five groups of teachers applied, making selection easy.

> *Staff development is one of the most important factors related to change in practice.* (Fullan 1982; Austrom et al. 1989)

Pilot teachers attended several in-service and curriculum development sessions and were given support to attend a regional mathematics conference. Although commitment among pilot teachers grew and interest within the department-head group increased, acceptance was still mixed. Another implementation tactic was added. A department-based staff development program with workshops on graphing calculators, cooperative learning, and data analysis was developed using some of the pilot teachers as presenters.

Since purchases of graphing calculators had to be spread over three years because of high cost and limited budgets, the initial plan was to extend the pilotlike approach to eighteen more teachers as a means of diffusing the innovation to the rest of the school system. (A similar "go only with those who are most interested" technique proved to build incentive when used by the author in implementing word-processing technology with English teachers in the suburban district. In that situation, one year's ambivalence was quickly eroded as the group of trusted colleagues from the few English departments that received considerable district support became the change agents in the second year and as school principals jockeyed for comparable resources.) The initial plan for the allocation and extension of the pilot, however, was revised when teachers at the cooperative learning sites gave cogent arguments for providing each student in a class with a graphing calculator and the added staff development proved more popular than anticipated. It looked as though every mathematics department would need at least one class set within two years.

It was at this juncture, mid-implementation, that the standard for the graphing calculator chosen by the urban district was changed. Unlike the commodity nature of the basic four-operation calculator (style and price may vary, but seldom the functions), the graphing calculator is still evolving. Shifting standards, which have sometimes plagued the implementation of computer technology, are unavoidable with the introduction of advanced calculators. The change in calculator standards was an agonizing decision for those coordinating the implementation, given the support system built for the first graphing calculator. This transition and the delay it caused might have made the implementation process more difficult, but the advantages of the new graphing calculator were clear, and to date all mathematics departments have indicated interest in completing in-service programs as soon as it is convenient. Fortunately, sufficient funds have been secured to meet this demand. In fact, as an experiment in a small school, every twelfth-grade student enrolled in mathematics, physics, or chemistry has been loaned a graphing calculator as standard equipment for school and home use.

Interestingly, several neighboring suburban districts that began in-service programs with graphing calculators before the urban district now have floundering implementations due, it seems, to the lack of targeted hardware funding for calculators. Unfortunately, senior district administrators sometimes view graphing calculators as a luxury or a fad and, failing to realize the fundamental changes this tool can have for the learning and teaching of mathematics, do not vigorously support the necessary funding.

CONCLUSIONS

Situations faced by "change agents" vary greatly. For effective implementation, strategies should be chosen that match the opportunities and obstacles of the innovation. The three vignettes discussed above offer examples of the diverse range of strategies that are available and at times necessary for success. The following list summarizes the strategies mentioned in this article:

- *Management*—analysis of obstacles and opportunities, collaborative implementation planning with school administrators, adequate funding for resources and calculators
- *Resources*—instructional and assessment guidelines, curriculum and resource development, textbook selection, assessment materials, selection and allocation of calculators

- *Staff Development*—on-site research, pilot testing, school-based in-service programs by teams of teachers
- *Community Involvement*—parent brochures, presentations to parent groups and school trustees

Our experiences suggest that when these strategies are thoughtfully and energetically employed, calculators can be made an integral part of school mathematics programs.

REFERENCES

Austrom, Liz, Roberta Kennard, Jo-Anne Naslund, and Patricia Shields. *Implementing Change: A Cooperative Approach.* Vancouver, B.C.: British Columbia Teachers Federation, 1989.

Beattie, Ian, and Walter Szetela. *Calculator Problem Solving Activities: Grades 3–8.* Toronto: Houghton Mifflin Canada, 1988.

Berman, Paul, and Milbrey McLaughlin. *Implementing and Sustaining Innovations.* Federal Programs Supporting Educational Change, vol. 8. Santa Monica, Calif.: Rand Corporation, 1978.

Fullan, Michael. *The Meaning of Educational Change.* Toronto: Ontario Institute for Studies in Education Press, 1982.

Hall, Gene E., and Susan F. Loucks. "Teacher Concerns as a Basis for Facilitating and Personalizing Staff Development." *Teacher's College Record* 80, no. 1 (1978): 36–53.

Hembree, Ray, and Donald J. Dessart. "Effects of Hand-held Calculators in Precollege Mathematics Education: A Meta-Analysis." *Journal for Research in Mathematics Education* 17 (March 1986): 83–99.

Price, Jack, and Jane D. Gawronski, eds. *Changing School Mathematics: A Responsive Process.* Reston, Va.: National Council of Teachers of Mathematics, 1981.

National Council of Teachers of Mathematics. *Agenda for Action.* Reston, Va.: The Council, 1980.

Super, Doug. *Calculators in the Elementary School: Information for Parents.* Vancouver, B.C.: British Columbia Teachers Federation, PSA Services, 1988.

Szetela, Walter, and Doug Super. "Calculators and Instruction in Problem Solving in Grade 7." *Journal for Research in Mathematics Education* 18 (May 1987): 215–29.

24

Statewide In-Service Programs on Calculators in Mathematics Teaching

George W. Bright
Patricia Lamphere
Virginia E. Usnick

During the current era of reform in mathematics education, mathematics teachers have special opportunities to restructure the ways they teach their subject matter. For this to happen simultaneously in many schools, however, there is a critical need for teacher in-service programs. In-service programs on the use of technology are an obvious case of need, since technologies are so new and virtually no teachers were exposed to technologies during their own learning of mathematics. This paper outlines the approach Texas used in developing a network of trainers who can help teachers learn how to use calculators in mathematics instruction in grades PK–10.

There are two obstacles to incorporating calculators into mathematics instruction. The first obstacle is convincing parents and administrators that calculators are legitimate tools to use to learn mathematics. Convincing people that *computers* enhance instruction does not seem to be a problem, but convincing them that *calculators* can also play this role seems more difficult, possibly in part because everyday uses of calculators generally involve only routine computation. Many people feel that calculators take the place of learning and that children will not need to think if calculators are being used. Actually, of course, the opposite is true; much thinking at conceptual and problem-solving levels is needed when calculators are used.

The work reported in this paper was supported by grants from the Dwight D. Eisenhower Education Act (National Programs Grant No. R168D80076 and Texas Education Agency Contract No. 89690301) and was completed while the authors were on the faculty of the University of Houston. All opinions expressed are those of the authors and do not necessarily reflect the positions of any government agency.

Providing a philosophical base for helping people change their thinking requires time and patience.

The second obstacle is educating teachers about appropriate uses of calculators in teaching mathematical concepts. Mathematics should be taught differently when calculators are in the classroom; for example, standard drill and practice exercises are no longer appropriate. Texas began to address this in-service need in 1986 by initiating the Mathematics Staff Development Project at the Texas Education Agency (TEA). The project is supported by the state's share of block-grant funds for the improvement of mathematics instruction from EESA, Title II, and more recently, Dwight D. Eisenhower programs.

MATHEMATICS STAFF DEVELOPMENT PROJECT

Beginning in 1986, the Mathematics Staff Development Project has funded the development of teacher in-service modules (typically twelve-hour short courses, carrying career ladder credit). These modules have focused on the use of manipulative materials or technologies, concept development, and problem-solving applications to enhance the teaching of the Texas Essential Elements of Mathematics (i.e., prescribed mathematics content for grades K–12). Each module covers one content strand (e.g., geometry) typically at one of three grade ranges (PK–2, 3–6, and 6–8), though some modules also deal with content in grades 9–12. The "overlap" of grade 6 in two contexts is due to the potential for grade 6 students to be in an elementary school or a middle school. Each module is unique, but all share the common purpose of helping teachers teach mathematics better through the use of manipulatives, technology, and problem solving. The expectation is that this improved teaching will result in improved student learning.

Each module was developed by a contractor chosen from among those who submitted formal proposals. To date, twenty-six modules have been developed, of which twenty focus primarily on the integration of manipulatives into instruction; the other five deal exclusively with technology (calculators and computers). The calculator modules use calculators capable of performing fraction operations and integer division (e.g., the Texas Instruments Math Explorer); the computer modules are typically written for both Apple and IBM. Calculator and computer modules for middle school were extended to cover grades 6–10 so that teachers of general mathematics courses would also have in-service materials to help them incorporate technology into instruction. Since four high school modules (algebra, geometry, precalculus, and prealgebra) address the use of technology, the technology modules did not need to be extended to those high school subjects. But it was felt that both the content in the algebra, geometry, and precalculus modules and many of the particular uses of technology were too sophisticated for most general mathematics students.

Once a module is written, the authors of the module train forty specialists selected by TEA from all parts of Texas; these specialists become the "designated trainers" for that module. The specialists have included teacher educators, in-service coordinators from the twenty regional Educational Service Centers in Texas, school district curriculum coordinators, and classroom teachers; at least one, and usually two, specialists come from each of the twenty regions. The designated trainers receive one to two days of training covering both the content of the module and techniques for delivering the module to teachers. Each trainer then receives a manual (including handouts for teachers, overhead transparencies, instructions for the delivery of the module, and background information) along with a kit of manipulatives (or access to software) sufficient to deliver the workshop to thirty to thirty-six teachers at a time. Each trainer makes a commitment to deliver a module at least three times. As a result of the project, module workshops have been given to thousands of teachers in Texas.

The plan of operation for the project allows information to be shared with many teachers throughout Texas in a systematic way while at the same time maintaining quality control over the substance of the workshops. Modules are typically a set of scripted activities that demonstrate the use of particular manipulatives or technologies that are known to be effective for illustrating and communicating mathematics concepts. Providing scripts (or script outlines) for trainers helps to standardize the balance between discussions of the mathematics and the pedagogy; the simultaneous training of the forty designated trainers for each module assures that teacher participants in subsequent workshops receive the same general exposure to the central mathematical ideas in the modules and to the use of manipulatives for teaching those ideas. It is emphasized to trainers that they should follow the framework of the scripts, though they are given flexibility to adapt the scripts and activities to their own personal styles and to the needs of any particular group of teachers.

CALCULATORS IN GRADES PK–10

One difference in pedagogy between the activities participants complete in the module and typical textbook instruction is the chance to explore many examples during the development of conceptual understanding. These examples can provide a strong experiential base from which concepts can be developed. Without an underlying conceptual base, typical textbook exercises are not likely to make much sense to students; we need to take advantage of calculators to help build that conceptual base.

The module developed for grades PK–2 (Schielack 1989) focuses on the use of calculators as tools for investigating patterns that build concepts of number, numeration, operations, relations and functions, and problem solving. The calculator is frequently used as a record-keeping device for results of work with manipulatives. The emphasis is on the mathematics being learned, with the technology serving as a medium to facilitate instruction. To extend students' experiences with concrete materials, calculators are often used in conjunction with manipulatives such as counters or base-ten blocks. Activities in the module also help teachers address concerns they may have regarding calculator use in instruction. Many teachers feel that the use of calculators inhibits students' learning of basic computational facts and skills. For example, in one activity, teachers are asked to list personal concerns about the use of calculators and then to list concerns that they believe other people have. These lists are compared and differences are discussed first within small groups and then in the whole group. As teachers work through the module, they get a sense for how calculators can be used to encourage conjecturing and critical-thinking skills through problem solving.

The module written for teachers of grades 3–6 (Dockweiler and Riley 1989) similarly attempts to lead teachers to the conclusion that the calculator is a useful device and has many applications to the teaching of mathematics. Major emphasis is given to concept development, problem solving, and reinforcement and maintenance activities. Participants use calculators and problem-solving processes to explore whole numbers, fractions and decimals, number theory, and probability and statistics. The activities are written to encourage discussion among the teachers, with carefully modeled integration of calculator use and mathematics content.

As a more detailed example of calculator training, we will look more closely at the module for grades 6–10 (Bright, Usnick, and Lamphere 1989). This module provides middle school and general mathematics teachers with calculator applications at levels appropriate for students dealing with prealgebra content. Since most middle and high school teachers have not personally studied mathematics with the help of calculators, the ways that mathematics is embodied in the module are likely to be unusual for teachers, even though most of the mathematics is familiar in other contexts. Workshop leaders have an opportunity during their presentation of the module to model for teachers how to handle a diversity of levels of understanding and confidence about both the technology and the particular embodiments of the mathematics while keeping everyone focused on the same mathematics. By the end of this module, teachers should have learned a lot about how to teach mathematics with calculators.

Content of the 6–10 Module

One important goal of middle school mathematics is to help students make transitions from elementary school thinking to high school thinking; that is, from concrete-based thinking to more abstract thinking. One major area in which such transition is needed is algebraic thinking (including work with fractions, ratio, and proportion). Much of the material in the workshop deals with these topics.

The 6–10 module is divided into seven sections: (1) Introduction to the Calculator (one hour), (2) Whole Numbers and Numeration (two hours), (3) Rational Numbers (two hours), (4) Number Theory (two hours), (5) Prealgebra (two hours), (6) Problem Solving (two hours), and (7) Closing (one hour). Part 1 introduces the fraction keys, so that participants who have little experience with a fraction calculator can begin to become comfortable with using the calculator. Participants who are already familiar with the calculator can skip this part and spend more time on the activities in the remainder of the module. Part 2 deals with place value, whole number operations, and number patterns. Part 3 includes the ordering of rational numbers, equivalent fractions, and estimations of fraction computations. Part 4 covers factors, multiples, and divisibility. Part 5 focuses on ratio and proportion, powers, equation solving, and complex fractions. Part 6 contains activities on simple and compound interest and problem-solving extensions of some of the previous activities. Part 7 gives closure to the entire module. These sections were organized to give approximately equal weight to whole number concepts, rational numbers, number theory, prealgebra, and problem solving. These topics are all approached through the unique powers of a fraction calculator to develop new ways of thinking about mathematical ideas that are fairly familiar, at least to teachers.

Sample Activities

To get a feel for the types of activities provided for teachers, two examples from the module are presented here.

Example 1. This example is from the whole number section. Participants are asked to experiment with the [10^n] key and then to play a game involving its use. (The number keyed in after pressing this key is interpreted by the calculator as the value of n, the exponent.) To develop notions of expanded notation in a calculator environment, participants are asked to make the number 3456 appear in the display, under the constraint of using the [10^n] key once for each digit. Keystrokes are recorded for each digit, and the number is written in expanded notation using exponential form. One keystroke sequence that would work is as follows:

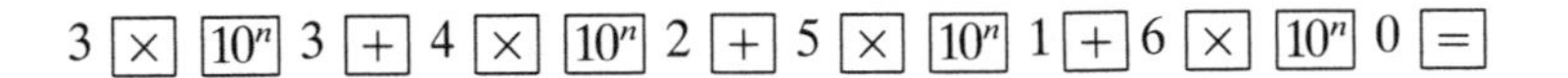

3 [×] [10^n] 3 [+] 4 [×] [10^n] 2 [+] 5 [×] [10^n] 1 [+] 6 [×] [10^n] 0 [=]

A discussion of the processes used by the participants is followed by more examples, as needed by any particular group of participants, in order to understand the relationship between the [10^n] key and place value. Then a game, called SPACE INVADERS (fig. 24.1), builds on this understanding of place value. Later in the module, this game is extended to decimal numbers. Keystroke sequences in that setting require the use of negative exponents.

Space Invaders

1. Clear the calculator display.
2. Enter the following number: 63418275. These eight digits are the "space invaders." In order to protect planet Earth, you must shoot them down one by one by changing each digit to 0.
3. Digits must be shot down in ascending order. That is, you must first eliminate the 1, then the 2, then the 3, and so on.
4. For each digit in turn, record a keystroke sequence below that includes [10^n]. Then carry out the keystroke sequence.
5. If the correct digit is removed, proceed to the next digit. If not, reenter the previous display and try again.

Digit	Keystroke Sequence	Number in Display
1	____________	63408275
2	____________	63408075
	.	
	.	
	.	
7	____________	8000
8	____________	0

Fig. 24.1

Example 2. This example deals with number theory. The fraction calculator simplifies fractions by removing the smallest prime factor from the numerator and the denominator with each press of the [Simp] key. This capability can be used to study the greatest common factor (GCF) of two numbers by entering one number as the numerator and one number as the denominator, simplifying the resulting fraction, and keeping track of the factors that are removed. The product of those factors is the GCF. A game (see fig. 24.2) was developed to let pairs of students investigate the GCF of pairs of numbers.

SUPERTRAINERS FOR THE CALCULATOR MODULE

The primary drawback of the plan for teacher in-service programs provided by the Mathematics Staff Development Project is that the trainers were not authorized to train more trainers; this problem seemed especially acute for the technology modules, since so few teachers know how to use technology in teaching mathematics and the subsequent demand for these workshops is high. The burden for providing in-service instruction to thou-

GCF Game

1. Decide who plays first. Play then alternates.
2. On your turn, your opponent chooses an uncovered number on the grid below.
3. You then choose a second uncovered number on the grid. Cover both choices.
4. Your score for the round is the GCF of the two numbers.
5. The winner is the player with the greatest cumulative score after five rounds.

25	60	45	15	10	80
48	64	36	24	65	99
27	16	42	81	75	25
200	300	500	600	800	900
360	480	640	550	270	120
144	625	525	648	864	468

Fig. 24.2

sands of teachers throughout Texas was squarely, and continuously, on the shoulders of the forty trainers. Further, the project plan did not address the role that administrators or school board members might play in the actual implementation of technology in teaching mathematics.

To help alleviate these problems, the University of Houston applied for and received funding from the national programs component of the Dwight D. Eisenhower Education Act to conduct a training project that would dramatically increase the number of trainers for the technology modules and would develop an in-service plan for school board members and school administrators, the support of whom seems clearly needed if we are to improve the quality of mathematics education. School board members in particular play a critical role in the implementation of curriculum innovations by providing the support necessary for teachers to actually deliver the innovations.

Evaluations of Training Sessions

The reactions of both supertrainers and ordinary trainers to the modules has been uniformly positive. One big hurdle in getting calculators used in classrooms seems to be in having teachers design activities for use with students. The modules furnish clear examples of activities that are effective with students; most activities for participants can be taken almost directly into classrooms. After trainers work through these examples, they seem to have an "aha" experience: "*Now* I know what to do with calculators!" Those trainers are then able to extend these examples into coherent curriculum plans that they can pass on to teachers.

Participants in the workshops delivered by trainers are often surprised at the mathematics content they learn through the activities. Comments such as "I didn't know that," "I wonder if this is always true," "I never thought of it that way before," or "What would happen if . . ." are often heard, with other teachers commenting, "Now kids will have to pay attention to how numbers relate to each other rather than just compute answers." Participants seem surprised that so much learning can take place so quietly when small groups are used. It is not unusual for the workshop leader to have to entice the small groups to come back together and *stop* sharing ideas in the small groups so that sharing can be done with the whole group. The excitement and enthusiasm generated by the teachers' active involvement in learning seem to be carried over into their classrooms. After their in-service training, many of the participants report back to trainers both that they used variations of the activities and that their students became very excited about learning.

IN CLOSING

The "training of trainers" model for the delivery of in-service programs to teachers in large geographic or heavily populated areas seems to be a powerful one, *provided that quality control is maintained over the substance of the in-service sessions.* The Mathematics Staff Development Project, which was initially conceived and implemented through the efforts of Cathy L. Seeley (director of mathematics) and her staff at the Texas Education Agency, has struck a good balance between specifying what will go on in in-service workshops and giving workshop trainers space for creativity. We recommend that other states, large school districts, and regional education service centers seriously consider using a similar model to furnish consistent in-service experiences for mathematics teachers. This approach gives teachers a common base of understanding and common vocabulary so that when they meet at professional conferences they can communicate better with each other about improving mathematics instruction. This enhanced level of communication among teachers is a significant positive force for improving mathematics education that will pay off for many years to come.

REFERENCES

Bright, George W., Virginia E. Usnick, and Patricia Lamphere. *Calculators for Grades 6–10.* Austin, Tex.: Texas Education Agency, 1989.

Dockweiler, Clarence J., and Alicia Gay Riley. *Calculators for Teachers of Grades 3–6.* Austin, Tex.: Texas Education Agency, 1989.

Dockweiler, Clarence J., and Jane F. Schielack. *Technology Inservice for School Administrators and School Board Members.* Austin, Tex.: Texas Education Agency, 1989.

Lamphere, Patricia, Virginia E. Usnick, and George W. Bright. *Technology Inservice for Mathematics Teachers.* Houston: University of Houston, 1989.

Romberg, Thomas A., E. Anne Zarinnia, and Steven R. Williams. *The Influence of Mandated Testing on Mathematics Instruction: Grade Eight Teachers' Perceptions.* Madison, Wis.: National Center for Research in Mathematical Sciences Education, 1989.

Schielack, Jane F. *Calculators/Computers: PreK–Grade 2.* Austin, Tex.: Texas Education Agency, 1989.

25

Introducing Calculators in Swedish Schools

Hans Brolin
Lars-Eric Björk

MANY people still believe that the primary purpose of school mathematics is to teach students how to do numerical calculations by hand. The primary purpose, however, must be to teach students *to understand and solve the problems they confront—*

- in everyday life;
- at work;
- in other school subjects;
- in the subject of mathematics itself.

Research in Swedish schools during the 1970s showed that students in traditional curricula were not acquiring those necessary problem-solving skills from instruction that emphasized drill in computational algorithms (Brolin 1990).

In short, we can say that the problem-solving process consists of the following steps (Polya 1945):

1. Formulate the problem.
2. Understand the problem and devise a plan for solution.
3. Carry out the plan.
4. Look back and check the results.

In the past, both for a lack of good calculating devices and in the belief that the calculations themselves provided comprehensive understanding, we devoted most of our time in mathematics education to step 3. For students, "solving a problem" is synonymous with the calculations involved in step 3.

THE ADVENT OF THE CALCULATOR

When the pocket calculator was introduced into Swedish schools in the mid-1970s, it was evident that the teaching of mathematics could undergo a major revolution. *The time spent on manual calculations could now be allotted to problem solving.*

Calculators were not introduced into schools by the National Board of Education or any other authority. It was, in fact, students in the secondary schools (aged sixteen to eighteen) who began to bring calculators to their science and technical courses. Teachers were uncertain what attitude to adopt and therefore turned to the National Board of Education for advice. Would it be permissible to use calculators in the classroom and during examinations? The teachers were allowed to use their own discretion as far as teacher-made examinations and classroom situations were concerned. However, during the examinations set by the National Board of Education, the use of calculators would not be permitted. Developments came fast, however, and by the beginning of the 1977–78 school year, calculators had gained acceptance—even for the board's central examinations.

THE POCKET CALCULATOR AND THE ARK PROJECT

In the spring of 1976, the Swedish National Board of Education appointed a group whose task would be to provide *an analysis of the use of the pocket calculator.* This task became known as the ARK project (Brolin 1987). The ARK project first needed to define the problems at hand. It decided to investigate *three aspects of calculator use* in grades 4–12 (ages ten to eighteen):

- The calculator as a computational tool
- The calculator as an instructional aid
- The curriculum changes arising from calculator use

The work was organized into ten subprojects, and many questions were asked. The general conclusions from the ARK project have been summarized in Björk and Brolin (1984). The most important subprojects were the RIMM and IAB projects, which studied the calculator as a computational tool in elementary schools.

Swedish students acquire their basic skills in algorithmic calculations in grades 4–6 (ages ten to twelve). Much of the time allotted to mathematics in these grades is devoted to teaching algorithms. Despite this fact, the results achieved by the students have long been below expectations. Thus, it was interesting to see what would happen if training in algorithms in these grades was reduced in favor of extended use of calculators for solving more complicated problems.

The RIMM project was established to examine this question. During the course of this project, standards of proficiency required in performing algorithms were lowered; for example, it was considered sufficient if the students were able to divide by single digits. In this way the investigators hoped to create wider scope for problem solving by calculators. Experimental material was developed and tested in grades 4–6 during the school years 1978–81. After this, the test material was revised, and a major trial took place in the period 1979–82.

It was extremely important to see the effect that a simultaneous reduction of time spent on algorithms and an increased emphasis on problem solving had on the following areas:

1. The nonalgorithmic basic skills
   - Number concepts (decimals, fractions)
   - Unit concept
   - The ability to choose the correct method of computation
   - The ability to interpret tables and diagrams
2. The ability to perform estimations and approximate calculations
3. The ability to calculate by hand (algorithms)
4. The ability to solve problems

A parallel subproject known as the IAB project was established to provide an instrument for measuring the above-mentioned skills and insights (af Ekenstam and Greger 1982). Under the auspices of the IAB project, tests were produced to permit measuring the nonalgorithmic skills of students in grades 4–6. By means of this instrument, the ARK project was given its first insight into what happens when the calculator is used and the amount of time devoted to algorithms is reduced. Those eight classes that took part in the RIMM project's major trials were tested in grade 6 in the spring of 1982 (Hedrén and Köhlin 1983). The results from the tests are summarized in table 25.1.

Table 25.1
Percent of Correct Answers

Skill area	Traditional classes	Experimental classes
Nonalgorithmic basic skills (49 questions)	63.5	72.0
Estimations (10 questions)	42.5	52.5
Algorithms (10 questions)	76.1	77.2
• Whole numbers	85.8	88.4
• Decimals	66.4	66.0
Mental arithmetic test (15 questions)	56.7	61.2

Table 25.2 gives a few problems from the tests and compares results of those classes that used the pocket calculator (experimental classes) to those that did not (traditional classes).

Table 25.2
Percent of Correct Answers by Item

Problem	Traditional classes	Experimental classes
Three pieces of cheese weigh 0.974 kg, 0.947 kg, and 1.050 kg. Circle the weight that is closest to 1 kg.	23	86
A cabbage costs 1.41 kr and weighs 0.94 kg. To find the price of 1 kg, which operation would you have to do? $1.41 + 0.94$ 1.41×0.94 $1.41/0.94$ $0.94/1.41$	41	54
Circle the best approximation. $1.02 \times 2.1 \approx$ 0.002 0.02 0.2 2 20	58	79
Approximately what percent of 85 kr is 44 kr? 15% 32% 41% 50% 85%	47	54
Perform the operation 23×6.	96	96
Perform the operation 43.5 / 15.	51	44

This investigation in grades 4–6 clearly indicates that the students who used the calculator—

- gained a better understanding of fundamental concepts;
- gained a better ability to choose the correct operation;
- gained a better proficiency in estimation and mental arithmetic;
- did not lose important basic skills in algorithmic calculations.

CONCLUSIONS OF THE ARK PROJECT

In 1983, the ARK project made the following recommendations:

Grades 4–6 (ages ten to twelve)

A large-scale trial use of calculators should be initiated at this level at the earliest possible moment. This would provide an invaluable opportunity to improve students' necessary skills as well as their level of comprehension. It is important that the standard of required basic skills be properly defined and that the students then receive thorough training.

Grades 7–9 (ages thirteen to fifteen)

At this level the calculator should function as a natural aid while previously acquired skills are kept fresh. A review of basic skills in estimation and in mental and written calculation should be carried out regularly. Calculators should be used in the assessment procedures.

Grades 10–12 (ages sixteen to eighteen)

Students in these grades should use calculators to perform the tedious calculations previously done by hand or with tables; the time gained should be spent on making mathematics more applicable to the real world, more experimental, more problem oriented, and more concept oriented.

DISCUSSION

What has happened since 1983? To what extent have the recommendations from the ARK project been fulfilled?

Grades 4–6

At this level the development has been very slow. In a recent survey (Ljung 1990) only 5 percent of the students in grade 5 gave an affirmative answer to this question: "Do you occasionally use a calculator at school?" Yet one-third of the students said that they used a calculator at home.

The main reason why the calculator is not taken advantage of in grades 4–6 seems to be the attitudes and beliefs of teachers. In these grades many teachers still see the teaching of algorithms as the primary function of mathematics courses. Consequently, they regard the use of calculators as a threat to their role as teachers. During the 1980s, teachers seemed to be more willing to discuss reasons why their students could not perform complicated algorithms as well as expected than to deal with the important question, "To what extent *should* students perform pencil-and-paper algorithms?"

During the last two years, a nationwide discussion of the aim and methods of mathematics, due to our students' poor performance on subtests from the Second International Study, has led many teachers to recognize the role of the calculator in school mathematics. It is hoped that the unnecessary fear of calculators among these teachers can be alleviated by gradual movement toward a course of study that emphasizes better understanding rather than drill and routine tasks. Also, teachers are encouraged to enhance the motivation of the lower-attaining students by letting them use the calculator as a numerical tool to solve problems.

Grades 7–9

Even at these grade levels the development of calculator use was slower than expected. However, the calculator is now used on most teacher-made examinations and on some of the tests given by the National Board of Education. Most textbooks have broadened their range of application problems and now try to teach students to make sensible choices among calculator, pencil and paper, and mental computation. Only a few textbooks still tell students specifically when to use a calculator. Estimation, handling significant digits, and doing problems with real data are now important topics in the textbooks. In short, the calculator has initiated useful discussions regarding the goals of school mathematics in the middle grades.

Grades 10–12

By the 1970s, all textbooks and all assessment and examination procedures were already based on the fact that students at this level would have scientific calculators of their own. The teachers' attitudes in these grades at all times been very positive. Relieved from the drudgery of complex calculations, the students can spend more time on problem solving and understanding concepts.

Curriculum changes at the beginning of the 1980s have allowed numerical approaches in calculus and more advanced methods in new topics of statistics to be studied.

FROM CALCULATORS TO COMPUTERS

Programmable calculators are now in frequent use, and for graphing and function studies mathematical workshop programs on microcomputers are used by many schools (Björk 1987). A large-scale trial called the ADM project is currently investigating the use of these new tools in mathematics education (Brolin 1990). The questions dealt with by the ARK project are generalized in this trial to computer use. Currently, the ADM project is dealing with—

- using the new tools as an aid for routine work;
- using the new tools as an instructional aid;
- the curriculum changes arising from the new tools.

The questions are now on other levels, but the answers are very much the same:

Reduce the routine work and teach for understanding.

REFERENCES

Björk, Lars-Eric. "Mathematics and the New Tools." In *Proceedings of the IFIP TC 3/WG 3.1 Working Conference on Informatics and the Teaching of Mathematics,* edited by David C. Johnson and Frank Loris, pp. 109–15. Amsterdam: North-Holland, 1987.

Björk, Lars-Eric, and Hans Brolin. *The Ark Project, a Report for the Period 1976–1983.* Stockholm: Liber Utbildningsförlaget, 1984.

Brolin, Hans. "Mathematics in Swedish Schools." In *Developments in School Mathematics Education around the World,* edited by Izaak Wirszup and Robert Streit, pp. 130–55. Reston, Va.: National Council of Teachers of Mathematics, 1987.

———. "The Role of Calculators and Computers in the Teaching of Mathematics." In *Developments in School Mathematics Education around the World,* vol. 2, edited by Izaak Wirszup and Robert Streit, pp. 189–209. Reston, Va.: National Council of Teachers of Mathematics, 1990.

af Ekenstam, Adolf, and Karl Greger. "Non-Algorithmic Basic Skills." *Journal für Mathematikdidaktik* 1 (1982): 19–44.

Hedrén, Rolf, and Kenneth Köhlin. *Miniräknaren på mellanstadiet.* Falun, Sweden: Department of Teacher Training, 1983.

Ljung, Bengt-Olov. *Matemstiken i nationell utvärdering.* Stockholm: Department of Teacher Training, PRIM-gruppen, 1990.

Polya, G. *How to Solve It.* Princeton, N.J.: Princeton University Press, 1945.

26

Part 6

Calculator Activities for the Classroom

CALCULATORS change what we can and must teach in the mathematics classroom, and they also change the ways that we can engage students in mathematics thinking and learning. The articles in this yearbook indicate the broad dimensions of the changes in goals and instructional approaches brought about by calculators. In this section we include an array of specific calculator-based activities for use in mathematics classrooms from primary school through secondary school and college.

Some of the activities presented below are games that can be played by pairs or groups of students independent of teacher guidance. Other activities are best run with a teacher leader.

ACTIVITY 1: FIVE STEPS TO ZERO

Doug Williams
Max Stephens

Intelligent use of a calculator requires well-developed number sense, including facile use of mental arithmetic requiring single-digit number facts. The following activity helps students develop this important understanding and skill.

Procedure: An activity for two players and one calculator

1. *Player A enters a three-digit number less than or equal to 900.*
2. *Player B must reduce the given number to 0 in at most five steps, using any of the four basic operations of arithmetic and a single-digit number at each step.*

For example, Player A might enter the number 703.
Player B might then work as follows:

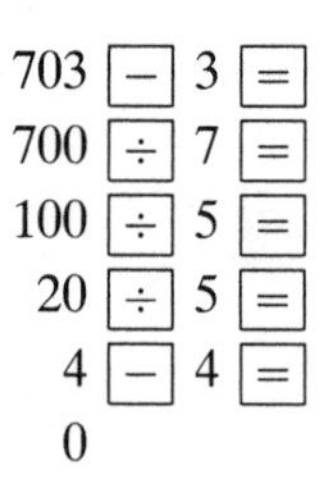

3. *To check results and study strategy after the game, players should record each move and the result as they go along.*

Clearly, success in this game depends on being able to use exact divisors. If one of the players happens to use a divisor that produces a decimal fraction display, this move is counted as a penalty turn and the player must return to the previous number displayed.

The particular value of this activity lies in the opportunities it offers children to discuss, compare, and improve their strategies. Students can be encouraged to reduce the number of moves made on some of their earlier attempts or to find numbers less than 900 that cannot be reduced to 0 in five steps. If students are using calculators with an algebraic operating system and integer arithmetic, the game can be extended to numbers between −900 and +900.

ACTIVITY 2: SQUARES

Doug Williams
Max Stephens

This activity can be introduced by a group discussion of the question, "If you are shown the square of a number, such as 196, what can you say about the original number?" This discussion should help students focus on the significance of the ones digit and help them estimate the target square root to the nearest ten.

Procedure: An activity for two players and one calculator

1. *Player A enters a number and calculates its square. Then the display is shown to Player B, who has to find the original number without using the calculator's* [√] *key.*
2. *Player B must tell Player A each number being tested. Player A then keeps a record of the attempts.*
3. *When Player B finds the required square root, players reverse roles.*
4. *After an agreed number of rounds, the player with the lowest total number of attempts is the winner.*

Students with less arithmetic knowledge and skill can be encouraged to begin with two-digit numbers and then extend to three-digit numbers. In the early years of school, before children know their basic arithmetic facts, the game can be played starting with a single-digit number, and students may prefer to declare a winner after each round rather than keep a total score over several rounds.

ACTIVITY 3: HOW MUCH SIMPLER CAN YOU GET?

George W. Bright
Virginia E. Usnick
Patricia Lamphere

Calculators that display rational numbers in both decimal and common fraction forms and convert directly from one form to the other provide a special opportunity to help students develop number sense. The following game, based on features of the Math Explorer calculator, can increase student knowledge of whole number factors and the equivalence of common fractions, decimals, and percents. Any other calculator with the same capabilities could be used, though there might be some minor modifications needed in the game rules.

Prior to playing this game, students need to be familiar with the calculator's fraction/decimal conversion key and the simplification key. On the Math Explorer, the simplification key removes the least prime common factor from the numerator and denominator with each press. The fraction/decimal conversion key "ignores" trailing zeros; for example, both 0.25 and 0.250 are converted to 25/100. Also, since the largest denominator allowable in the display is 1000, a decimal like 0.2345 will not be converted to a common fraction by this calculator.

Procedure: Any number of students can play, taking turns in the following sequence of game steps.

1. *The player whose turn it is enters a decimal fraction of at most three digits and presses the key that converts that decimal to a common fraction.*
2. *The player presses the simplify key until the displayed fraction is in simplest form. The player's score depends on the number of times the simplify key must be pressed to reach simplest form:*

zero presses	*1 point*
one press	*2 points*
two presses	*4 points*
three presses	*8 points*
four presses	*16 points*
five presses	*32 points*

3. *Record both the entered decimal and its simplest common fraction equivalent. Once a decimal has been used by any player, it may not be used again by any other player.*
4. *The winner is the player with the most points after a predetermined number of rounds.*

Clever students will find a variety of strategic insights that lead to maximum point totals. For instance, since decimal denominators are powers of 10 (on the Explorer, limited to $1000 = 2^3 5^3$), numerators with factors like 5^3 or 2^3, without any factors of the other prime, will be "three press" decimals. Whenever 2 and 5 appear together as factors, they will create a trailing 0, which will be ignored in the conversion to a common fraction form. The game has some obvious variations—like limiting entries to two decimal places to focus on fraction equivalents of various percents.

(Preparation of this activity was partially funded by the Texas Education Agency, through the Education for Economic Security Act (EESA, Title II) and the Eisenhower Education Act. All opinions expressed are those of the authors and do not necessarily reflect the positions of the Texas Education Agency or any other government agency.)

ACTIVITY 4: BULL'S EYE

Virginia E. Usnick
Patricia Lamphere
George W. Bright

When calculators are used to perform tedious computations, students are able to focus on other mathematical concepts and skills. The following game gives students opportunities to practice and refine their estimation skills while working with rational numbers.

Procedure: An activity for several players and one calculator—preferably a calculator that can use mixed-numeral inputs and convert common fractions to decimals. The youngest player goes first, and play then proceeds in a clockwise fashion. At each turn the player does the following:

1. *Spins whole number and fraction spinners like those below and uses those results to form a mixed number A, and then repeats the process to form a second mixed number, B.*

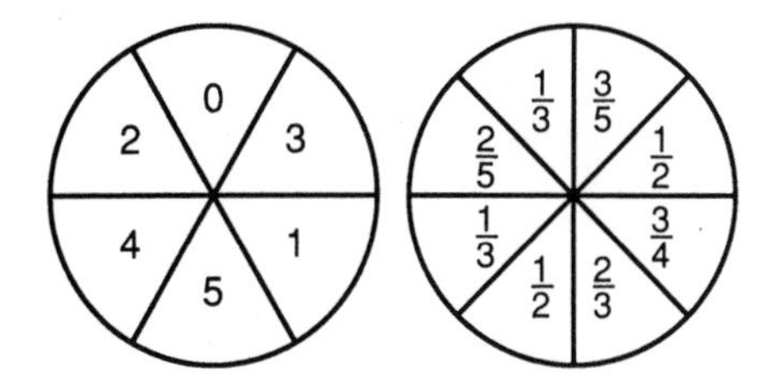

2. *Spins an operation spinner (like the one shown here) to select an operation.*

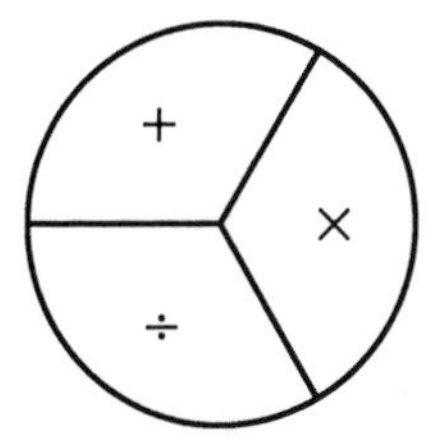

3. *Estimates the answer for "A operation B" and records that estimate.*
4. *Uses the calculator to determine the exact value of "A operation B."*
5. *Compares the estimate and the exact value by division, "estimate/exact," rounding this value to the nearest hundredth.*
6. *Locates the comparison quotient on the scoring number line shown below.*

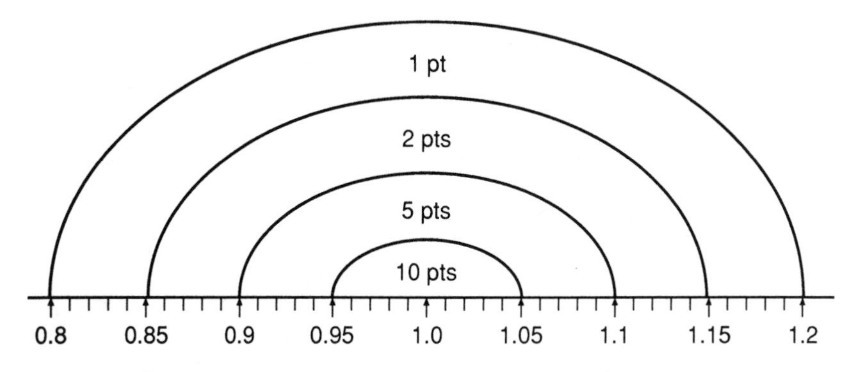

7. *The winner is the player with the highest total after a predetermined number of rounds.*

Variations on the basic Bull's Eye game can include using only whole numbers for younger students or using decimals instead of common fractions on the fraction spinner.

(Preparation of this activity was partially funded by the Texas Education Agency, through the Education for Economic Security Act (EESA, Title II) and the Eisenhower Education Act. All opinions expressed are those of the authors and do not necessarily reflect the positions of the Texas Education Agency or any other government agency.)

ACTIVITY 5: PRESS ON!

C. J. Dockweiler
Jane F. Schielack

The standard method of reading and writing whole numbers and decimal fractions proceeds from left to right. The entry of numbers on a calculator follows the same order, but the displayed number changes with each successive entry. The following simple calculator activity emphasizes place value by tracing the effect of successive digit entries. The activity can be done in a teacher-directed whole-class lesson or as a lab activity for small groups of students.

Procedure: The activity requires one calculator; if done with the whole class, an overhead display would be useful.

1. *Clear the calculator and press 1. Then ask:*
   Q: What number is displayed?
2. *Press 2 and ask:*
   Q: What happened to the 1 and what is its new value?
   Q: What number is in the display now?
3. *Press 3 and ask:*
   Q: What happened to the 1? To the 2? What are their new values?
   Q: What number is in the display now?
4. *Press 4 and ask:*
   Q: What happened to the 1? To the 2? To the 3? What are their new values?
   Q: What number is in the display now?

There are several obvious variations on this activity. Entering a different sequence of digits, like 6 7 8 9 or 3 5 2 8, will help highlight the place-value effects. Beginning with a press of the decimal point and a single digit, followed by other digits, is a useful contrast to the effect of entering successive digits in a whole number. Of course, entering longer sequences of digits will reinforce other place values.

ACTIVITY 6: GUESSTIMATE: HIGH/LOW

C. J. Dockweiler
Jane F. Schielack

Several kinds of low-cost calculators capable of representing and operating on rational numbers in common fraction form are now available. Those calculators can help students gain valuable number and operation sense with fractions.

Procedure: This investigation can be pursued by individuals or small groups of students who have access to a calculator that operates on common fractions.

1. *Begin with a set of four numbers, a, b, c, and d.*
2. *Form two fractions using the given numbers as numerators and denominators and use your calculator to find the sum of the two fractions:*

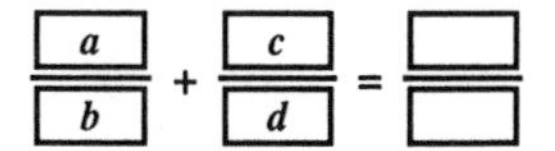

3. *Write other possible pairs of fractions using the four given numbers and find the pair that gives the greatest sum.*
4. *Repeat steps 1–3 with several different sets of four numbers, looking for a pattern that tells how any given numbers should be arranged to find the greatest sum.*

This fraction activity can be varied in several ways to provide different challenges. For example, try a set of four numbers in which two or more are the same. Try a set of numbers that are not all less than 10 or another in which one of the numbers is 0. Ask students to find the placement of numbers into fraction form that will produce the *smallest* sum. In response to any answer students give—either for a specific set of given numbers or for a general rule for maximizing or minimizing the sum—be sure to ask for a justification of the answer.

ACTIVITY 7: WIPE OUT—REFINED

Martha H. Hopkins

Calculators can be used effectively to teach and reinforce the concepts and notational conventions of place-value numeration. Wipe Out (Judd

1979), one of the best-known activities with that objective, challenges students to change single digits of a numeral to 0 by single calculator operations. For instance, students might enter 54628 and be asked to wipe out the 4 (which can be done by subtracting 4000).

When we played the original version of Wipe Out with children, we found that they could play successfully without knowing the place values of the digits being removed. The following revision of the original rules awards maximum points to those players who can correctly identify the meaning of the wiped-out digit.

Procedure: Each student player should have a calculator.

1. *The teacher begins play by writing a multidigit number, with no two digits the same (like 54628), on the chalkboard, overhead projector, or overhead projected calculator. Students enter the same number on their calculators.*
2. *The teacher identifies one of the digits (for example, the 6) to be wiped out, and students silently try to change only that digit to 0 using only one operation.*
3. *The teacher chooses at random one student to report the procedure used. Points are scored as follows:*
   - *Student reports only the operation and the digit to be wiped out (for example, "minus six"); teacher earns 2 points.*
   - *Student reports only the operation and digits entered (for example, "minus six, zero, zero"); teacher earns 1 point.*
   - *Student reports the number subtracted (for example, "minus six hundred"); class earns 2 points.*
4. *Play continues for a predetermined number of rounds. The team with the highest score wins.*

The difficulty of this game can be varied by altering the number of digits allowed in the original number or by using decimal fractions. After several practice sessions with the teacher, students can be encouraged to play this game in pairs or in teams.

ACTIVITY 8: CALCULATORS AND READING NUMBERS

Valerie J. Malkus

Often students can use their calculators to produce very large numbers but cannot read those numbers in a way that shows an understanding of place value. The following activity is a way to use calculators to practice place-value skills and help teachers check students' work very quickly.

Procedure: Each student or group of students working together will need a calculator.

1. *Provide each individual or group of students with a series of numbers written in words. For example:*

   (a) Two thousand three hundred six and five hundredths

   (b) Seven thousand

   (c) Six hundred five and three hundredths

   (d) Two hundred four thousand

   (e) Eight hundred and eight hundredths

2. *Students are to enter each number into their calculator, pressing* [+] *after each number. They should also record the calculator display at each step.*
3. *Students report the calculator display at the end of their entries. The teacher can quickly check to see if the work is correct and quite possibly spot the nature of any error. For example, the display for the numbers given in (1) should be 214711.16.*

Variations of this activity include making the numbers to be entered simpler for students who have less number knowledge or presenting the numbers in expanded numeral form. For example:

(*a*) one hundred thirty-four	(*a*) 2 tens 6 ones
(*b*) three hundred six	(*b*) 7 tens
(*c*) seven hundred fifty	(*c*) 4 tens 3 ones
Sum 1190	Sum 139

When using very large numbers and decimals together, some caution is required because calculators can display only limited numbers of digits, and they often round decimals to provide the displayed number.

ACTIVITY 9: GEOMETRIC SEQUENCES, SQUARING, AND SQUARE ROOTS

Alfinio Flores

The constant [K], square [x^2], and square root [$\sqrt{x}$] keys on calculators can be very useful tools for an investigation of important sequences. The following examples suggest several of the most interesting patterns that students can discover.

Procedure: Individuals or groups of students need calculators with [K], [x^2], and [$\sqrt{x}$] *keys to explore these situations.*

1. *Geometric Sequences—Enter a number c as a constant factor and observe the pattern of results in c, c^2, c^3, c^4, . . . by using the constant key of your calculator. Then try a variety of positive and negative values for c and see if you can find patterns that allow you to predict the behavior of such a sequence from properties of the common factor c.*

   Students should notice that for $c > 1$, the sequence grows without bound; that for $-1 < c < 1$, the sequence approaches 0 as its limit; and that for $c < -1$, the sequence alternates positive and negative terms, and absolute values grow without bound.

2. *Squaring—Enter a number c and press the* [x^2] *key repeatedly. Since each term in the sequence is the square of its predecessor, this process is producing the sequence c, c^2, c^4, c^8, c^{16}, ... Again, try a variety of positive and negative values for c and see if you can find patterns that allow you to predict the behavior of such a sequence from properties of the common factor c.*

   Students should note that the behavior of this sequence is similar to the geometric sequences in (1), except that convergence and divergence occur much faster and negative values of c do not produce alternating signs in the sequence. The rapid convergence and divergence can be explained by noting that the squaring sequence is a subset of the geometric sequence.

3. *Square Roots—If you begin by entering a positive number c and repeatedly press the* [$\sqrt{x}$] *key, you will be producing terms in the sequence c, $c^{1/2}$, $c^{1/4}$, $c^{1/8}$, ... Once again, try a variety of positive and negative values for c and see if you can find patterns that allow you to predict the behavior of such a sequence from properties of the common factor c.*

   Students are likely to be surprised that if $c > 1$, the sequence approaches a limit of 1 from above. If $0 < c < 1$, the sequence approaches the limit of 1 from below. As with the squaring operation, these sequences approach their limits very quickly.

Looking ahead to more advanced topics, students with scientific calculators might explore the effects of repeatedly pressing the [sin] or [cos] keys.

ACTIVITY 10: THREE APPROACHES TO THE GOLDEN RATIO

Alfinio Flores

Many interesting number sequences are produced by repeating a set of calculator operations to obtain the successive terms. The set of keystrokes define an iterative process that produces each term from its predecessor(s). There are several such iterative algorithms that lead, in the limit, to the familiar golden ratio.

Procedure: This investigation can be carried out by an individual or group of students and a single calculator. The first stage requires a [$\sqrt{x}$] *key, the second a* [$1/x$] *key, and the third a calculator that handles rational numbers in common fraction form.*

1. *Calculate terms in the sequence* $\sqrt{1}, \sqrt{1 + \sqrt{1}}, \sqrt{1 + \sqrt{1 + \sqrt{1}}}$, *. . . . To obtain a term in this sequence, add 1 to the previous term and take the square root of that sum:* $S_n = \sqrt{1 + S_{n-1}}$. *Repeat the same series of keystrokes to obtain successive terms until the display stabilizes. The result should be something close to* $r = 1.618033989$. *Next, square the limit r to obtain* $r^2 = 2.618033989$ *and notice the apparent relations* $r^2 = 1 + r$ *and* $r = \sqrt{1 + r}$.
2. *Next, calculate terms in the sequence defined by* $S_n = 1 + 1/S_{n-1}$. *Each term is obtained by adding 1 to the reciprocal of the previous term. The terms of this sequence should also approach* $r = 1.618033989$. *Taking the reciprocal of r will give* $1/r = 0.618033989$, *suggesting the intriguing relationship* $r = 1 + 1/r$, *which is equivalent to the relations obtained in (1). The positive root of the equation is the limit of both sequences and is known as the* ***golden ratio.***
3. *Using a calculator that handles fractions without converting automatically to decimal form, express the terms of the sequence obtained in (2) as common fractions. This should produce the sequence* $\frac{1}{1}, \frac{2}{1}, \frac{3}{2}, \frac{5}{3}, \frac{8}{5}, \frac{13}{8}, \frac{21}{13}, \frac{34}{21}, \frac{55}{34}, \ldots$ *. It is a surprise that the denominators and numerators are successive Fibonacci numbers! The Fibonacci sequence 1, 1, 2, 3, 5, 8, 13, 21, 34, 55, 89, 144, 233, 377, 610 is defined by the recurrence relation* $F_n = F_{n-1} + F_{n-2}$. *Thus, to approximate the* ***golden ratio,*** *we divide a term of the Fibonacci sequence by the preceding term to get* $S_n = F_n/F_{n-1}$.

Clearly, these investigations of Fibonacci numbers and various iterative algorithms for producing approximations to the golden ratio could not be carried out nearly as easily without the use of the calculator!

ACTIVITY 11: CONTINUED FRACTIONS

Lida Garrett McDowell

Every rational number can be expressed as a finite continued fraction, and every irrational number can be approximated by such a fraction. For example, the continued fraction expansion for $\frac{67}{24}$ can be developed as follows:

$$\begin{aligned}
\frac{67}{24} &= 2 + \frac{19}{24} \\
&= 2 + \cfrac{1}{\frac{24}{19}} \\
&= 2 + \cfrac{1}{1 + \frac{5}{19}} \\
&= 2 + \cfrac{1}{1 + \cfrac{1}{\frac{19}{5}}} \\
&= 2 + \cfrac{1}{1 + \cfrac{1}{3 + \frac{4}{5}}} \\
&= 2 + \cfrac{1}{1 + \cfrac{1}{3 + \cfrac{1}{\frac{5}{4}}}} \\
&= 2 + \cfrac{1}{1 + \cfrac{1}{3 + \cfrac{1}{1 + \frac{1}{4}}}}
\end{aligned}$$

This continued fraction is written in condensed notation as $<2,1,3,1,4>$.

There is a simple calculator algorithm that helps produce these continued fractions. It is easiest to describe the algorithm by showing how it is applied to a specific example:

$$\frac{67}{24} \approx 2.7916667$$

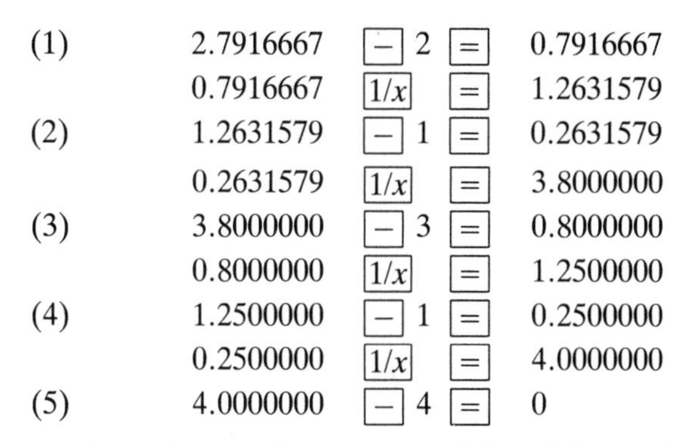

Step	Display	Keys	Result
(1)	2.7916667	[−] 2 [=]	0.7916667
	0.7916667	[1/x] [=]	1.2631579
(2)	1.2631579	[−] 1 [=]	0.2631579
	0.2631579	[1/x] [=]	3.8000000
(3)	3.8000000	[−] 3 [=]	0.8000000
	0.8000000	[1/x] [=]	1.2500000
(4)	1.2500000	[−] 1 [=]	0.2500000
	0.2500000	[1/x] [=]	4.0000000
(5)	4.0000000	[−] 4 [=]	0

This procedure gives students a powerful tool for exploring properties of continued fractions. The following suggestions will generate interesting patterns:

1. *Find the continued fraction expansion of* $\frac{24}{67}$ *and compare it to the expansion of* $\frac{67}{24}$.
2. *Find the expansion of* $-\frac{7}{5} = -2 + \frac{3}{5}$.
3. *Find an approximating expansion for* $\sqrt{2}$, $\sqrt{3}$, $\sqrt{5}$, *and so on by using the* [$\sqrt{x}$] *key to get started. Remember that* $\sqrt{2} \approx 1.4142136$.
4. *Find the continued fraction expansion of the golden ratio* $\frac{\sqrt{5}+1}{2}$.
5. *Find an approximating expansion for* π *by using the* [π] *key to get started. Note that* $3 + \frac{1}{7} = \frac{22}{7}$ *is the rational approximation we use for* π. *"Better" rational approximations can be obtained by continuing the expansion.*

In general, continued fraction expansions of real numbers give "good" approximations for those numbers. The subject has rich potential for investigations that deepen students' understanding of numbers and their fractional representations.

REFERENCE

Judd, Wallace. "Instructional Games with Calculators." In *Calculators: Readings from the Arithmetic Teacher and the Mathematics Teacher,* edited by Bruce C. Burt. Reston, Va.: National Council of Teachers of Mathematics, 1979.

Index